BLACK TIDES

My father? The sun
Is my father;
The earth is my mother
Who nourishes me,
And on her bosom
I will recline

———————————————————

Tecumseh, leader of the Shawnees

BLACK TIDES

Miles O. Hayes

UNIVERSITY OF TEXAS PRESS, AUSTIN

To my father, *Norman Eustace Hayes* (1901–1989),
who showed me, by example, how to love the earth
and gave me total freedom to find my own way.

and

To *Jacqui Michel, Erich Gundlach, Ed Owens,
John Robinson, Dave Kennedy,* and *Jerry Galt,*
my oil-spill heroes, who kept the faith
through all those *post-spill grand blue funks.*

Quotes from *Cold Mountain* by Han-Shan, transl. Burton Watson, ©
1970 by Columbia University Press. Used with permission of Columbia
University Press.

Requests for permission to reproduce material from this work should be
sent to Permissions, University of Texas Press, P.O. Box 7819, Austin,
TX 78713-7819.

∞ The paper used in this book meets the minimum requirements of
ANSI/NISO Z39.48-1992 (R1997) (Permanence of Paper).

Library of Congress Cataloging-in-Publication Data

Hayes, Miles O.
  Black tides / Miles O. Hayes.
    p.   cm.
  Includes bibliographical references.
    ISBN 978-0-292-73124-0
    1. Oil spills—Environmental aspects. 2. Oil spills—Manage-
ment. 3. Coastal ecology. 4. Hayes, Miles O. I. Title.
QH545.O5H39   2000
363.738'2—dc21   99-20518

Contents

Foreword: A Major Oil Spill Here

Walking around in oil up to my knees on the shoreline of southern Chile in August 1975 was a sobering experience for a nature lover like me, particularly because I had just spent several weeks working along the essentially untouched coastline of south-central Alaska. Eventually, though, my practical, scientific nature took over, and I saw that there was a job to be done that I was uniquely qualified for, first academically and later on as a business person. Some of my students and business partners and I have been trying to understand how to respond to such incidents ever since.

I started keeping a fairly detailed diary a year or so before that, so I have a pretty good record of what went on at the oil spill in Chile and the ones that followed. This book is built from a mosaic of those diary notes, plus a little reminiscing about the more distant past. The incidents reported are all true. In a few instances, the time sequence has been changed slightly in order to smooth the telling of the story. Also, where appropriate, the names of certain individuals have been changed in order to protect their privacy.

I don't have a clue why I took all those notes; however, over time, pieces of the mosaic seemed to fit together as if predestined—pieces such as living in the rain forest in Alaska, the Black Tide of La Coruña, Spain, the plane crash, and on and on. The mosaic took on a near-final form as I walked the oil-soaked mud flats of the coastline of Saudi Arabia after the Gulf War oil spills of 1991. But I'm getting ahead of myself.

25 December 1978—
On the beaches of Puerto Rico

The young Coast Guard petty officer, sweat pouring down his face, waved his arms emphatically while trying to direct traffic at the intersection of two dirt roads several yards behind the beach. Long lines of cars were backed up on both of the roads, with half of the impatient drivers honking on their horns. The traffic was a complete snarl on this Christmas holiday weekend on the north shore of Puerto Rico.

"There's been a major oil spill here!" he shouted his frustration, somehow thinking that this proclamation would clear the traffic. Then the cleanup of the No. 6 fuel oil that had spilled out of the barge *Peck Slip* and was impacting the beaches could begin in earnest.

We finally reached him and showed him our badges. He waved us on by. "Which way to the beach with the heavy oil?" I asked as we passed him. He pointed to the left, and we drove out of the traffic, following the ruts over the dunes and down to the beach where the cleanup workers were still gathering.

By then, we were doing this for a living: going to oil spills whenever and wherever they occurred if the U.S. Coast Guard needed some scientific support from our client, the National Oceanic and Atmospheric Administration (NOAA), to plan a response to the spill. We had been on worse assignments than Puerto Rico in December. This spill was fairly early in the game for us. We had only been at it for three and a half years, and we were still green and excited, glad to get the chance to go to the spill and make our contribution. So what if it was Christmas? We still had a lot to learn. We saw heavily oiled mangroves for the first time at that spill, and I was impressed by the transport of the dark black oil back out to sea by currents in the rip cells that developed all along the exposed sand beaches.

The science we pursue is called "coastal geomorphology," which means study of the shape, or form, of the coast, including how it has evolved. About thirty-five years ago, while still a graduate student at the University of Texas, I began my first scientific research project on an ocean shoreline, the coastal bend area of South Texas. I finished my dissertation on the geomorphological effects of hurricanes on the Texas coast and journeyed on to teaching positions at the universities of Massachusetts and South Carolina. At those schools, I was joined by an army of graduate students as we conducted studies of shorelines as far away as northern Alaska, southern Chile, and Kuwait.

When we started, understanding how the coast was made was almost virgin territory, at least in the United States. Hypotheses on how barrier islands evolve, how the tides and waves shape the coast, and how the storms change things were there for the alert observer to deduce, assuming enough examples had been seen and that the global processes had been properly accounted for. We were using what Comet (1996)[1] referred to as "one of Aristotle's two forms of logical inference"—namely, inductive reasoning from the observation of a generalized pattern or distribution in order to develop a principle or law. In other words, we started with a large number of more or less random observations, not with a detailed data set on a specific topic. However, we did eventually try to figure out how to collect meaningful data sets to verify or disprove those deduced principles or laws.

The practical reasons for making such observations usually centered on environmental issues such as beach erosion and other forms of habitat destruction. The most meaningful observations were made by being there—we went by boat, by plane, by four-wheel drive vehicles, and by walking. We spent long hours in the sun and the wind and the rain, working out ways to make our observations more and more disciplined and useful.

Most of the observers I worked with were called to this field by some elemental communication with the Great Mystery or the Spirit Father (or whatever the people that beat us here—by about ten thousand years—called it). Over the years, we all came to love that part of the earth where the waves and the tides shape the land. It seemed as if we almost became a part of it ourselves.

The methods we used to pursue our dream of understanding how the coast was made are no longer in vogue, a kind of scientific future shock for people like me. The new dreamers who want to understand this phenomena will not be doing it by foot, by car, by plane, or by boat. They will be doing it with remotely sensed images and with computer models that churn out mathematical solutions to their theories. I just hope that the new searchers will somehow be able to feel it, smell it, and touch it the way we did.

Through all of our wanderings we noticed how the most remote and unmodified areas of the coastlines we studied were being rapidly encroached upon by humans and how the most populated areas were abused more and more over time. We tried to deal with these abuses each in our own way. Leaving the bigger, and no doubt much more significant, issues such as ground water depletion, global warming, and sea level rise to others, my immediate associates and I have focused on a problem more tangible and approachable, the *impacts of oil spills.*

It all started at the end of the earth, along the shoreline of Chile in the Strait of Magellan in the summer of 1975, and almost ended for me fifteen years later on a rock off Mitkof Island in southeast Alaska in the summer of 1990.

Acknowledgments

A special thanks to my wife, Jacqueline Michel, for her continued encouragement to write this book and for her critique of the manuscript. Ed Levine and Alan Mearns of the National Oceanic and Atmospheric Administration (NOAA) made many helpful suggestions for improving the text. Shannon Davies, sponsoring editor of UT Press, Mandy Woods, and Sheri Englund also improved the manuscript with their insightful comments. At RPI, Dot Zaino finalized the word processing, and Joe Holmes produced the graphics.

These stories owe their origin to the many great teachers and great students I have been associated with over the years. Thinking back to Oakley High School, the teachings of Mrs. Blankenship, Mr. Wells, and Coach Burchfield come to mind. As discussed in the text, in graduate school I was inspired by the work of Professors James C. Brice of Washington University and Robert L. Folk of the University of Texas. Graduate students too numerous to mention individually carried out the work and taught me much more than I ever taught them. The same can be said for my fellow scientists at RPI.

Much of the financial support for the research described here was provided by the U.S. Coast Guard and NOAA. Contract monitors at NOAA who guided us included John Robinson, Dave Kennedy, and Bob Pavia. The work on the Gulf War spill was supported in part by the Marine Spill Response Corporation, under the overview of Don Aurand.

BLACK TIDES

PART I

HEAVEN'S

DOOR

The Crash

Sunday, 12 August 1990 —
On a rock off Mitkof Island, Alaska

I shivered as the sun went behind the high cirrus clouds for the third or fourth time since I had been there. I reached behind me with my right hand, stretching the muscles to the limit against the sharp pain near my right shoulder blade. But I had already plucked up all the grass within the reach of that arm, so I came up empty handed. My left arm lay just on the high-tide line, where no grass was growing.

Then I placed my right hand on the grass piled on my bare chest and pressed down, absorbing the warmth the brown fibers had taken from the sun while it was out. I shivered again, my teeth chattering as I lifted my wristwatch in front of my eyes. It was 10:20 A.M., dead low tide.

We had been down for fifty minutes. It would be another two and a half hours before JM and her pilot would miss us and start searching.

"Todd is dead," I said to myself as I remembered the start of the day.

JM and I were walking along the road in Petersburg, Alaska, heading for downtown when Todd Hatfield, our pilot who had spent the night with friends, stopped for us in a cab. We had breakfast with him at Margie's Restaurant. He was very quiet as JM and I babbled away about the science of the project. This was to be the last day of our field work in southeast Alaska. We would be approximately halfway through mapping the shoreline of the area for environmental sensitivity (to oil spills), which meant we had met the specs of our 1990 contract with the National Oceanic and Atmospheric Administration (NOAA). We were having a good time, concluding that this beautiful, diverse setting was probably our favorite part of Alaska. It was also relaxing to be away from the politics of the *Exxon Valdez* oil-spill site for a change.

We were working with two fixed-wing planes. Todd would be my pilot in a yellow-and-white Cessna 185 on floats. We were scheduled to depart at 8:30 A.M. and work the four-hour window around low tide. The tides control all scheduling of this type of work, which is essentially mapping the intertidal zone, especially in southeast Alaska where the vertical range of the tide exceeds twenty feet at times.

Todd perked up when we started discussing the mechanical aspects of the mapping, asking a number of questions about typical air speeds, flap settings, and altitudes. He had flown me over the mountains to Petersburg from Ketchikan the afternoon before, after I had completed my low-tide mapping session in that area with another pilot. JM had already been mapping in the Petersburg area for several days. Todd had not flown a mapping mission in Alaska, but he told us about flying some state biologists in Florida who were mapping plant assemblages in the Everglades. He wondered aloud how the biologists were able to locate themselves on their maps.

"Yeah. It's tough down there," I responded. "Everything looks the same. Very flat!"

After concluding this leisurely discussion and leaving Margie's, we walked down the long pier to the float-plane dock. It was a sparkling clear day, and the sun was shining over our shoulders and reflecting off the white feathers of the hundreds of glaucous-winged gulls that circled over the harbor. I felt the chill of a light breeze, at least ten or fifteen degrees cooler than it had been at that time of day for several days, and wondered if I should put on my jacket. I didn't.

The right pontoon on our plane had sunk a bit overnight, so Todd began pumping the water out of it while JM and her pilot prepared to leave. I waved good-bye to her as they taxied out into the open harbor past myriad fishing vessels, some still anchored and some moving about in what appeared to be random motion from my vantage point. Their float plane, also a Cessna 185, lifted off promptly at 8:30 A.M., headed south.

We organized the interior of Todd's plane, and I climbed into the left rear seat, directly behind the pilot's seat. Todd was looking for the bolts to put the right rear seat back in place. We had removed it in order to pack gear back there for the trip over from Ketchikan the afternoon before. JM and I had unpacked most of the gear upon my arrival in Petersburg.

"Let's leave it like this," I told Todd. "I can lay my pack and maps on the floor, and, you're such a big guy, I can put my right leg over there."

He climbed into the left front seat and adjusted it, leaving barely enough room for my left leg, which I propped against the back of his seat. He was around six feet four inches tall and weighed about 220 pounds, I guessed, and he was built like a tight end. It never occurred to me to ask him if he had ever played football, but I found out later that he played music and wrote poetry. He was twenty-seven years old.

"This is fine, I can see out of both sides okay," I told him, as I adjusted the seat and pulled the life vest out from under it a little, noting the yellow color, while he started the engine and filed a flight plan.

After takeoff, we started mapping south along the Wrangell Narrows. We were moving rather slowly, probably 75–85 miles per hour. Because of where I was sitting, I could not see the instrument panel so I had to guess how fast we were flying. I thought about checking with him on the air speed, but was distracted by the changing landscape, and didn't think of it again. I had asked him to maintain an altitude of around five hundred feet.

As we started up Blind Slough, the flight service called to say that an ELT (emergency locator transponder) was going off somewhere in our vicinity. Todd passed the word that we should keep an eye out for any boats or planes in trouble.

When we reached the southwest tip of Mitkof Island at the south entrance to Wrangell Narrows, we turned east to continue mapping the shoreline. Shortly thereafter, Todd reported to the flight service that we would soon be approaching the south side of Blind Slough, which occupies a fault zone that cuts through the center of the island, and that we would continue mapping all the way around Mitkof.

The southernmost tip of Mitkof was on a map different from the one I was working with. JM, who was mapping to the south of us, had that map with her and was scheduled to map that small segment of Mitkof on her return to Petersburg. As we came back onto my map, I couldn't tell exactly where we were, so I asked Todd to turn back so I could check out the area again. He seemed to be concerned about making the turn over the open water and said he wanted to go behind a hill which lay a bit east of a small rocky embayment to our left. The hill was snug against the high, steeply sloping main body of the island. I wasn't paying close attention to our flight path as we approached the hill because I had managed to locate on my map the small bay, which had a complex mixture of tidal flats, rock outcrops, and mussel banks in it.

I began to map the features in the bay, continuing until it was hidden behind the small hill as we passed between it and the steep mountain. In order to pass around the hill, Todd turned the plane sharply to the left, with the left wing tilted down about fifteen to twenty degrees. As we passed back out to sea, a sudden jolt of strong turbulence tilted the wings into a vertical position. In the absence of the lift normally provided by the wings, the plane headed straight down, engine first, toward the ground.

During this cataclysm, Todd didn't say anything that I could hear; he had the engine on full throttle the whole time. It felt like a strong force was pushing on the top of the plane, as if it were an object being swung around in a circle on the end of a string. But the radius of the circle was too large for the plane to pull out before impact.

I can divide my thoughts during that brief five-hundred-foot trip to the ground at the speed of over a hundred miles per hour into three distinct segments, during which everything seemed to be moving in slow motion. Throughout the first segment, I just thought we had hit some bad turbulence but that Todd would be able to pull out of the dive, so I was not overly concerned, just surprised. However, during the second segment, as we continued downward and I was looking straight ahead, which is to say straight down at the water, I realized that we might crash, and I was vicariously flying the plane with Todd, mentally pulling back on the steering mechanism. Indeed, during that second segment, we did seem to pull out a bit, but then the plane tipped back up to vertical again.

When that happened, I was sure we were going to crash, and I said loudly, "My God, we are going to die!"

During the third segment, that short instant of time as we passed through the last hundred feet or so, I bent over to the side with my head facing the right rear window. I grasped my right leg with my right hand and pulled my body down, covering my head with the other hand. Apparently, I did not move my legs. My left leg remained pressed against the back of Todd's seat, and my right leg was stretched out under the copilot's seat, which had been pushed forward as far as it would go.

We hit in about ten or twelve feet of water, with the plane tilted down to the left at about a sixty-degree angle. I learned later that the left wing was broken off, and both of the pontoons were fractured on impact. Gasoline quickly leaked into the plane, no doubt as a result of the broken wing and its ruptured fuel tank.[2]

The first thing I remember about the impact was an abrupt noise like an explosion and the splash of water that covered the windows. Then a kind of high frequency, muted sense of pain was all around me. My body had been thrown up tight against Todd's seat as my seat broke off its bolts and scooted forward, and I think I skinned my rear end in the process. My right foot was jammed up under the right front seat, which had the instrument panel pushed into it. Although my seat broke off its base, the seat belt held, and it was pulling me tightly sideways as I lay stretched out on my back. I remember saltwater being in the plane; I could taste it. There was also a fairly strong smell of AVGAS, though I made little mental note of it and didn't think about it again for hours.

The plane then tilted slowly backward and upward to nearly a horizontal position, floating on the badly mangled pontoons. I pulled myself slightly upright and looked out the right window.

Instantly, I began talking loudly to myself, swearing profusely, and saying things like "We have crashed this blankety-blank plane," "I have to get out of here," and so forth.[3]

For a short time I refused to accept the fact that this had really happened to me, but then I heard a sound from the seat against my left shoulder. It was Todd, breathing out two sighs. After struggling with the seat belt for some time and then finally untangling myself from it and freeing both of my legs, I pulled my upper body up with my arms so I could look over at him. He was still clutching the wheel, and the instrument panel was tight up against his chest. His head was cocked upward and slightly to the side. It didn't quite register completely, but somehow I knew that he was dead. I learned later that he had broken his neck upon impact.

After reaching that conclusion, I suddenly felt an overwhelming sense of helplessness.

A heavy, heavy weight pushed me back down, and it felt like I was sitting on fire. While moving my upper body up and down, I had glimpsed strands of muscle hanging out of a huge gash in my left shin, no doubt cut when I lunged against the back of Todd's seat. I also noticed that I couldn't put any weight on my right leg, so both legs were out of commission.

And after having made all of those discoveries, I felt a certain reluctance to continue participating in that scene. "This is not happening to me! This is not really happening to me!" I thought. I couldn't quite come to grips with the idea, it had happened so quickly.

No doubt primarily because of the shock of Todd's death, what followed was a short period of time when my mind seemed to wander into a completely different arena. The exact details of this interlude escape me now, but my thoughts probably went something like this:

> I can't be here alone in this plane with . . . Todd, he's dead.
> Surely, I don't have to get out of here and lie over there on the bank like some forgotten refugee. I'm the infamous M. O. Hayes, self-proclaimed world's greatest coastal geologist. I've been to the South Pole and the North Pole and all around the world, at least that's what I told my high school class at our twenty-fifth reunion a few years back. In '76, Cowboy, Chris, and me flew the entire Alaskan coast. Hey, I took two thousand pictures. And in '75, Gayle, Chris,

Janie, and me crashed in the ocean up there by the Malaspina Glacier, and the plane kept running right up onto the beach. I've been in turbulence so bad that my head bounced off the roof of the plane, and Cowboy, he had me flying the plane upside down. Crap, I've done this job a thousand times.

And then I kept saying over and over, "Todd is dead! Todd's dead!"

The searing pain under my butt focused me again, and I reluctantly lifted up and put my hand inside my trousers, expecting to come up with some very bad news. I couldn't believe it when I saw that the retrieved hand contained only clean, clear water that was now beginning to fill the interior of the plane as it slowly sank.

I reached for the right door, and the next thing I remember I was outside the plane, balancing on the pontoon with my left foot, while hanging onto the strut with both arms.[4] Blood was spurting out of the gash on my left shin, sprinkling down on the sinking pontoon. The plane was tilted to the left, probably because the left pontoon hit the bottom first and with the most force and was, consequently, the most damaged in the crash.

During the entire period between when I discovered Todd was dead and the moment I was standing on the pontoon, it was as if two separate people were participating in that episode. One was a casual, though careful, observer of the scene who had moved outside of the injured geologist's body, so he could have a better observation post, I presume. The second man, the one with the real problems, was a struggling fifty-five-year-old who stood hanging onto the right strut of the plane, his shirt tail hanging out. As he stared in disbelief, the Prowalker shoe came off the foot he was holding up in the air and bounced off the pontoon, taking a quick dive of twelve feet or so to the bottom. Without the shoe, he could observe clearly why he couldn't stand on that foot. All the bones in the middle part of it had been formed into a teepee shape and protruded out of the place where the ankle used to be. The top of the teepee still had a cover of stretched skin.

He had been nearsighted from an early age and had lost his glasses in the crash; as he looked around, the coastline that he had just recently been mapping took on the appearance of an impressionist painting. The missing glasses were of little consequence to him. In fact, he hardly thought about them, because the colors he saw, though somewhat blurred, had meaning. He recognized the colors of the biozones of the intertidal zone—white for the barnacle zone, yellow-green for the popweed zone (*Fucus* sp.), and blue for the mussels (*Mytilus edulis*). In addition to being distributed in a distinct band along the base of the rocky edge of the water, the blue mussels also occurred as a number of isolated banks scattered about the bay, about three or four of which protruded out of the water over to the right of the plane. As the injured geologist was thinking about the mussels, the disinterested observer noted that the old guy was into science again. The fun was over, so to speak, and thus the two entities suddenly merged and weren't separated again during that crisis.

The maneuvering to get out of the plane had apparently been quite exhausting because I remember thinking, "I know I will have to swim over there to those mussel banks pretty soon, but I think I will just rest here for a little while."

The instant I had that thought, the plane started to sink much more rapidly. I quickly glanced to my left, just in time to see Todd's head disappear under the water, and then I started to swim as the water reached my waist. The nearest mussel bank, which was a few feet in diameter and stuck up out of the water only a foot or two, was probably fifteen to twenty yards away. After choosing the direction in which I would swim, I put my face in the water and stroked with my arms, not looking up until I felt the edge of the mussels with my hands.

As soon as I reached the mussel bank, I pulled myself onto it and sat down on the razor-sharp shells, which didn't exactly ease the pain I had been feeling in my rear end. Once again, I thought something terrible was wrong under my bottom; I therefore reached to check it out further. As before, no blood was in evidence as far as I could see. In addition, the bleeding in my left shin had stopped, the wound presumably having been cauterized by the cold water. When the realization of how cold the water was set in, my teeth started to chatter, and I rubbed my chest to stimulate the circulation.[5]

Soon after checking out my bottom, I once more began to doubt whether the crash had actually happened. I thought that maybe, hopefully, I was dreaming. "Time to wake up now!" I pleaded with myself, as I pinched my arms and slapped myself on the face. No such luck. Couldn't wake up, no matter what I tried.

For the first time, I looked at my watch, which, miraculously, was still on my wrist. It was 9:26 A.M. I guessed we had been down five or six minutes. I knew the ELT on the plane was under water and assumed that it was not working. I also knew that another ELT was going off somewhere in our vicinity, and as far as I could tell, no one was looking for it in any serious fashion. Needless to say, I didn't think I would be rescued in response to our ELT signal. I wasn't. (I guess I could question why the Federal Aviation Administration [FAA] even bothers to require these things to be in the planes, but I won't.)

Therefore, I thought there were two possibilities of my being rescued: a plane flying over or a boat passing by would spot the wreck—the yellow-and-white tail and the end of the left wing were still sticking out above the water; or JM and her pilot would get back to Petersburg around 1:00 P.M., discover we had not returned, and start looking for us. And since they didn't know precisely where we were, they would have to look a long time before they found us.[6]

Once more, I carefully looked around the area for a route of escape. In about forty-five minutes it would be low tide. When the tide turned, this mussel bank I was sitting on would be covered with water within a matter of minutes. Besides, I had to get out of that cold water somehow! Because I couldn't stand up, only my upper torso was actually out of the water and had any chance of drying out.

I had earlier made a quick assessment of my wounds. I didn't know how it could be done, especially repairing the hole in my shin, but I was sure our good old American medical profession could make me almost as good as new, even if my foot looked like a teepee held sideways. Thank God the bleeding in my left shin had stopped, and, of course, every once in a while I did wonder why my rear end was on fire. Nonetheless, with that confident overview completed, I didn't worry about much of anything from then on except getting out of the water and drying out.

I knew that I had to get off the mussel bank soon, so I looked over to my left where a rock island, or peninsula, sloped down into the water at a rather high angle. That rock, in addition to being the closest one to me, was also the highest in the area (about twenty-five feet high). The intertidal part of the rock contained the usual distinct biozones along its side, and I could see some terrestrial grasses on the top, which assured me that the top of the rock would be out of the water at high tide. The rock was fifty or sixty yards away from where I sat. A shallow tide pool over some mudflats that contained a few scattered mussel banks lay between me and the rock. To somehow get on top of the highest point in the vicinity of the wrecked plane, as well as above the high-tide line, seemed like a pretty good idea to me, so that was where I headed.

I pushed off the mussel bank and guided myself with my hands toward the rock island, floating in the shallow water over the flat, which was only three to five feet deep. The real fun started when I reached the base of the rock. I had twenty vertical feet to go to the high-tide line up the steeply sloping rock surface, which was covered with razor-sharp shells in the barnacle and mussel zones, and mushy, slimy algae in the

Fucus zone. As I pulled myself up the side of the rock, it seemed as if I were losing strength on every pull, and it didn't help any when I slipped back down the slope as I hit a particularly slick patch of algae. Sometimes I lay on my stomach and sometimes I turned and scooted on my back. The mussels and barnacles cut into my hands in a hundred places as I gripped the rocky surface.

I was quite cold and shaking as I very slowly worked my way up the rock. I continued talking loudly to myself, occasionally saying "Todd is dead!" and other lamentations. I never thought in terms of whether or not I would survive. In fact, I think I was surprisingly confident and calculating about the whole thing. However, I was very concerned about hypothermia; therefore, I started remembering the video that I had viewed on that subject while taking the HAZWOPER safety training sponsored by Exxon back in April. That training was part of the preparation for the summer field work at the *Exxon Valdez* oil-spill site. If someone had given me a quiz on that hypothermia video the night before the crash, I probably would have scored about 40 percent. However, as I was crawling slowly up the slick rock and replaying the most minute details of the hypothermia video in my mind, I would have scored 100 percent on any questions asked. I knew what I was going to do when I got to the top of the rock.

I scooted up onto the upper surface of the rock while lying on my back and found a little low spot above the high-tide line within which my body snugly fit. When I reached that position, the only clothes I had on were a long-sleeved field shirt, an undershirt, long pants, some undershorts, and a pair of gym socks. Everything that I had in my trousers pockets was gone, including my wallet, which I usually kept in my right rear pocket. When I started to take off my field shirt, I noticed that my clip-on polarizing sunglasses were still in the left pocket, having been put there for a fishing trip the previous day. (I don't wear sunglasses when I'm mapping because I have to be able to distinguish among the true colors of the sediments in order to make accurate calls on the habitat classification.) The clip-ons were the only thing, except the bare minimum of clothes and my wristwatch, that had managed to stick with me all the way to the top of the rock. When I pulled out the clip-ons, I was so mad that they were all I had with me, considering how absolutely useless they were, that I threw them as far as I could back down the rock.

After that little episode, I quickly went into my anti-hypothermia routine. Remembering that the most critical areas for heat loss from the body are from the top of the head, the chest area, and the private parts, I removed both of the wet shirts, carefully wrung them out and spread them out on the rock behind me, hoping they would dry out in the sun. Then I pulled up all the grass that I could reach from where I lay and piled it on my naked chest. When it was out, the sun, which was shining directly into my face, warmed up the grass. I unbuttoned my trousers and pushed them down as far as possible, being careful not to disturb my broken legs, and grasped my private parts, warming them with my left hand as best I could. I used my right hand to press down on the grass on my chest when the sun went behind the clouds.

The shivering and teeth chattering gradually subsided as the sun warmed me. I reckoned that I would be in for a long wait, at least four hours before JM and her pilot could find me, best case. I was completely exhausted and didn't move to get any more grass. At one point, I started to pray, as I had been for a while as I was crawling up the rock, but I felt so hypocritical about it that I soon stopped. I can't remember exactly what I said. I had some serious doubts about what prayer really meant. I think I said that if I were rescued, I would be ethically pure, although I thought I was trying to be that already—whatever that meant. It was just a brief thing. Nonetheless, it was pretty obvious that someone or something was doing a good job of looking out for me. Maybe it was my mother's prayers; she said she prayed for me every day, especially since I had quit going to church. I had quit that practice twenty-five years before the crash so she had been praying for me a long time.

As the time passed, I didn't move again. I was in serious pain. My bottom was killing me, and I had no idea why. My right foot was also hurting badly.

Time dragged by. A couple of airplanes passed over in the vicinity, but I never actually saw them. I played a game of guessing just how much time had passed within certain time intervals, usually three to five minutes. I got to where I could guess the time within ten seconds or so. No, I wasn't counting the seconds. I was glad to have the watch because it gave me something to do.

At about 11:25 A.M., I heard a motor out over the water, which meant a boat was in the vicinity. Then I saw a fuzzy white object round the point to the east of me and head in toward the little bay. I started waving my white undershirt while they were still far out, but they didn't see me. They were headed for the tail of the Cessna 185 with the broken floats. As the approaching vessel, which I reckoned to be about a twenty-five to thirty-foot sport-fishing boat, came closer, it passed out of sight behind the rock I was on. Although I had continued to wave the shirt, they had not seen me. I could still see the tail of the plane, but the boat was hidden from sight.

I started to panic, thinking "What if they just think it is an old wreck, turn around, and head back out to sea?"

The boat stayed out of my view behind the rock for what seemed to be an eternity, but then, finally, I saw it nose right up to the tail of the plane. I could see a man and a woman on deck.

"Hey! Hey!" I yelled at the top of my lungs, continuing to frantically wave the shirt.

They finally looked my way.

He yelled back at me, "What happened? Are you all right?"

"Yes, I think so. I appear to have two broken legs, but the pilot is dead."

"Do you need anything?"

"I sure could use a blanket."

Then I heard them calling someone on their radio.

He moved the boat around to a better spot than the one I had used to climb up the rock. The tide had come up quite a bit in the meantime, which also made the climb easier.

He walked up to me and introduced himself. His name was Jamie Debore, and he was about thirty-five years old. He was doing a little salmon fishing with his wife on a nice Sunday afternoon. After laying a piece of blue canvas over me (he didn't have any blankets), he took off his sweatshirt and wrapped my foot with it and a piece of plastic he had carried up from the boat.

He was yelling back and forth with his wife, who was still on the boat, about who else to call on the radio and giving other instructions. Then he said to me, "The paramedics are on their way from Wrangell. They should be here in about fifteen minutes." Sounded pretty good to me.

And then, as an afterthought, he added, "In a little while there will be more people here looking after you than you could ever imagine. The salmon are running, and there are a lot of people out on the water (who were probably listening in on the radio messages)."

"Anything else I can do?" he asked.

I was starting to shiver again. The canvas wasn't quite doing it. "Well, I'm still a little cold," I said, and he took the shirt off his back and gave it to me. I quickly put it on. I never felt anything so warm in my life. After that, he had to go back to the boat to get another shirt and jacket.

As Jamie had predicted, it was only a few minutes before more boats arrived and people began walking up the rock to me. One of the first to get there had something in his hand. "These yours?" he asked, as he handed me the clip-on sunglasses that I had thrown away. "Oh yeah," I said. I took them and placed them over my eyes because the sun was out again.

I told them about Todd, who had been under the water about two and a half hours by then. I had been trying, unsuccessfully, not to think about him.

"Here come the paramedics," Jamie announced.

I looked up and saw a float plane pull right up to the base of the little island we were on. A young man in his late twenties came bounding up the rock. He was thin, with blondish hair, and, as I remember, had a narrow mustache. I was struck by how much he reminded me of one of my graduate students, Tom Moslow, in his younger days. How could I help but feel comfortable with this guy? "What's your name?" he asked.

"Miles."

"My name is Brian. Don't worry, we are going to get you fixed up here," he assured me.

I lay back and closed my eyes, which were hidden under the clip-ons, as Brian and his associate checked my blood pressure, felt all over my body for injuries, checked the movement in my toes, and so forth. Then Brian started cutting off my trousers and the rest of my clothes, destroying Jamie's shirt in the process. I sent Jamie a replacement shirt later. It was a small price to pay someone for saving your life, no?

After Brian had removed all my clothing and covered me with blankets, he said, "Now I'm going to have to take your watch."

"Don't lose it. It cost me eight bucks," I countered.

As soon as I said that, I was startled by laughter coming from a lot more than two or three guys. I opened my eyes, removed the clip-ons, and saw a ring of approximately twenty men standing all around me. The sun was shining at an angle to them, and several of them cast shadows across me in the brilliant sunlight. I was lying bundled in the warm blankets, surrounded by a halo of exquisite light! At least it looked exquisite to me. Where did they all come from?

That watch lasted another three years.

Brian was talking back and forth on the radio with his associates at the hospital in Wrangell. I particularly noted the deep, resonant voice of one man who was apparently the head doctor. "Caucasian male, mid-fifties," Brian reported. "No. The pilot is presumed to be dead. What? He's under about fifteen feet of water." And so on.

"Damn, how did he know I was that old!" I thought. "Hey, I'm in good shape now. When I went fishing yesterday morning, I walked all the way up the mountain to that lake without stopping once. And I was wearing my hip waders. How far was that, four miles?"

I didn't stop to think how long it would take me to get back into shape after this adventure, but it would be a very long time, and I might never be able to repeat that mountain-climbing feat again.

There was a lot of discussion about how they were going to transport me forty miles over the open water to Wrangell. Brian was ready to take me in his float plane, but he talked to someone on the radio who countermanded his plan. He wasn't too happy about that. He finally walked over and told me that a U.S. Coast Guard H-3 helicopter was on its way over from Sitka, and that it would be a while before they could get there.

I reminded Brian to have someone call the flying service in Petersburg and tell JM that I was okay. She and her pilot would be getting back pretty soon.

At about 12:30 P.M., I heard the old familiar flt-flt-flt-flt-flt of the helicopter blades as it crossed over the hill and headed in toward the bay where we were. As soon as he heard it, Brian and several of the men put me on a stretcher, and Brian had to cinch me down tight. Being on the stretcher was extremely painful, primarily for my rear end, which had received no medical attention. But it surely was hurting in that position.

Meanwhile, the helicopter had landed down by the edge of the water on the tidal flat. "Hurry, Hurry!" someone yelled up from the flat, "The wheels are sinking in the mud!"

About eight people picked me up and raised me over their heads, and then we proceeded down the landward side of the rock that I had spent my last three hours on. I was somewhat concerned that they might slip, and we would all go tumbling down the rock. I held on tight to two hands, one on either side of the stretcher. One of them belonged to Brian. It was a bumpy ride down, and when we reached the tidal flat, they tilted the stretcher at about a forty-five-degree angle and shoved me feet first up into the waiting hands of the Coast Guard helicopter crew members. I had flown in numerous Coast Guard helicopters on our various mapping missions all over the country and the world, but this was the first time that I would be doing it as the target of a rescue mission.

I looked back down at Brian and said, "Brian, are you going with us in the helo?"

He said, "Don't worry, buddy, I'll be with you the whole time," which was something I was mighty glad to hear.

Then Brian climbed in beside me and said, "We'll be in Wrangell in ten minutes," as he handed me some ear plugs.

I had been on oxygen since Brian had arrived and was occasionally getting the shakes. He kept assuring me that I was doing great and seemed delighted that I had sensitivity in my feet and could move my toes.

At 1:00 P.M., they rolled me into the small emergency room of the Wrangell Hospital. I remember seeing numerous faces peering at me as they rolled me along. I was a real curiosity, somebody who had survived one of those airplane crashes out in the bush. At that time, my main concern was that I was incredibly thirsty. I remembered that I was just about to open a soft drink shortly before the plane crashed, and now, three and a half hours later, I was famished for a drink. However, I couldn't have one because they were not sure what operations I would have to undergo. I did manage to coerce one of the nurses into feeding me tiny pieces of crushed ice from time to time.

I eventually saw the doctor who I had been listening to on the radio. He was wearing Levi's and a red-checked flannel shirt for his Sunday afternoon duty. He was a tall, well-built, rugged-looking man of about my age. As I said before, he had a resonant and kind voice. First he stuck needles into my toes, which made me jump and yelp. So much for the kindness bit. Then they x-rayed my lower extremities and probed my body looking for other problems. Because the treatment I required was too complex for the Wrangell facilities, I would have to go to the Ketchikan Hospital, which was a bigger unit.

After completing the examination, the doctor told me kindly, for sure this time, "Well, Miles, looks like you are going to be all right. You're a real tough guy."

In response, I tried to say "But I wish I could have helped Todd." However, I think it came out something like, "Todd, he isn't here." And I thought the rest, "He's still in the plane." I couldn't finish. I looked away from the doctor and the faces of the nurses all around the table, and for the first time, the tears came.

Outside the hospital, they put me on a board stretcher and strapped me in real tight, pressing my unattended-to-butt against the hard board. This caused excruciating pain, the worst of the whole ordeal.

Then they put me in an ambulance to take me back to the helicopter for the trip over the mountains to Ketchikan, but the ambulance's motor would not start. It was very hot and stuffy inside the ambulance because it had been sitting in the sun. While they were trying to decide what to do, I told Brian that I had to get out of there and that I couldn't go on with my butt pressed against that board. He seemed a little disgusted with me. After all, wasn't this a minor inconvenience considering what I had just been through? But one of the female nurses convinced him that it was too hot and claustrophobic inside the ambulance, and thank God, they decided to put three pillows under my back so my butt would not be pressed so hard against that board. After that, the stretcher became barely tolerable.

JM was waiting for me inside the Coast Guard helicopter, wearing an orange Coast Guard Mustang suit. She had heard about the crash about 1:00 P.M. and the Temsco people in Petersburg had flown her to Wrangell in a helicopter. The Coast Guard personnel made her stay in her seat with the seat belt fastened, so she could not get near me.

At the time we lifted off, I still had not had any medication, and I was in considerable pain and felt groggy. But shortly thereafter, as we flew along in the helicopter noise, I looked over at JM and wrote "I love you" in the air with my finger. She looked puzzled at first, but when I did it again, she smiled and waved.

Brian seemed happy that I had gotten over the panic that began in the ambulance. We were going to go up to five thousand feet for the trip across the highest mountains, so they had to adjust the inflatable cast on my foot to allow for the changing altitudes. After making one of the changes, Brian grinned down at me and said, "You know what they call these things, don't you?" I shook my head no. "A bucket of ten thousand bolts that has no business flying."

I was thinking, "The fear of flying will no doubt catch up with me later, but right now, I surely am happy to be flying along in this bucket of bolts and not lying over there on that rock freezing to death!"

The Patient in Room 07

When the helicopter landed in Ketchikan, I was rushed into an ambulance, JM and Brian with me, and we drove onto the waiting ferry they had held up just for me. In Ketchikan, the air strip is carved into the side of the mountain across a deep channel opposite where the town is located, so one has to take a ferry to go back and forth to the airport.

After crossing the narrow channel, we bumped off the ferry, and soon I was talking with two young doctors in the small emergency room at the Ketchikan Hospital, telling them about the crash. One was a pilot himself, so he asked the most questions. The other was a surgeon, who was waiting to operate on me. He looked just like Duncan FitzGerald, another one of my graduate students. Hmmm. How did all of my graduate students get into the medical profession? I thought they were all coastal geologists!

The Duncan-look-alike surgeon introduced himself as we shook hands, "Dr. Bendowski."

"B-E-N-D-O-W-S-K-I," I spelled out the name slowly.

"Hey, most people have trouble with the name," he said as they wheeled me down the hall. Dr. Bendowski actually worked in a big military hospital in Fairbanks, and he was in Ketchikan doing a little moonlighting on his vacation so he could make enough money to buy himself a truck.

I had already said good-bye to Brian, who was on his way back to Wrangell. I was still extremely thirsty, but I was headed straight for the operating room. I looked at the clock above the door as we went inside, 6:30 P.M., exactly nine hours and five minutes since the crash.

I continued my conversation with Dr. Bendowski. "I had lots of Polish graduate students at the universities of Massachusetts and South Carolina. The name is always spelled just like it is pronounced."

I mused over those students as the medical staff was getting ready to administer the anesthetics. First, there was Ray Kaczorowski, who was by then a top executive in a mineral exploration company in Houston. He had been with me at both universities. And let's don't forget Danny Domeracki, former head of our company's Boulder office. You'll be hearing a lot more about him later. And Carl Yastremski. No? Right, he is a Hall of Fame baseball player, not one of my students, but I was living in Massachusetts in 1967, the year he won the triple crown and led the Red Sox to the World Series. What a year!

And then, before I knew it, I was fast asleep.

My lucky streak was holding. Dr. Bendowski did an excellent job, turning the teepee hanging off my right leg back into a normal-looking foot. The operation involved putting pins in the five major fractures in my right foot and putting both that foot and my left leg, which had also sustained a fracture, into casts, among other things. When I woke up in the recovery room, I asked for something to drink. That was the best Diet Pepsi I ever had!

A short time later, I was rolled into the room that I would occupy, with assorted roommates, for the next ten days. It was room 07, appropriate for one of the luckiest people in Alaska on that day.

JM was waiting for me in room 07, and it seemed that she hardly ever left it for those ten days, except when I was asleep. She slept in a chair right beside me that they had forced in between the two beds. I will never forget how she looked after me all that time. I guess it must be love. It was just a couple of months short of fifteen years earlier that she had first started putting up with me.

The next day was a pretty busy day despite JM's concerted efforts to let me get some rest. There were many phone calls. Danny Domeracki had called long into the night before, and we were finally able to talk the next morning, which was something we had done almost every day, except for when I was in the woods fishing, for the past eight years. He was still holding down the fort, such as it was, in Boulder.

Somebody showed me the article about the crash in the *Ketchikan Daily News*, which had a headline on page one that read "Temsco Crash Kills One." Parts of the story follow:

Todd Hatfield, 27, a Temsco pilot, died Sunday when his plane crashed into the water off the east point of Station Island outside Wrangell. His passenger, Miles Hayes, 55, was medivacked to Ketchikan General Hospital, where he underwent surgery, according to Alaska State Troopers . . . Wrangell Search and Rescue and emergency medical technicians responded as did a trooper tactical dive team. Hatfield's body was recovered and his family in Lincoln, Neb.

was notified. . . . The Federal Aviation Administration and the National Transportation Safety Board are investigating.

The first visitor I had was another Temsco pilot, Doug McCart, who had been my pilot for the early part of the work in the Ketchikan area. Doug, a big burly guy dressed in typical Alaskan field clothes, gave me a huge bear hug. "I'm so sorry," he said, "I should have been there."

I had known Doug for a long time because he had flown a couple of other mapping missions in Alaska for me, the last being in Bristol Bay only three or four years back. He's a pretty religious guy and Saturday was the Sabbath for his family, so rather than wait for him, I went ahead with Todd. Doug and his wife, Delgina, came to see me every day I was in the hospital, bringing goodies to eat and other gifts each visit. You meet the best people in Alaska. Too bad some of them are no longer with us.

That afternoon, the FAA investigators paid me the first of two calls and stayed a long time asking numerous questions. They apologized for staying so long, but explained that it wasn't very often that they had somebody to interview after a crash like that. I don't know what they wrote up as the reason for the crash, but I'm sure the words "stall out at a speed too slow at an altitude too low" must have been in the report somewhere.

Somebody finally discovered why my rear end had been bothering me. It had third-degree burns all over it. The AVGAS that came into the plane right after the crash had caused the burns. I guess the burns would have been more severe on other parts of my body if I hadn't taken most of my clothes off. My backside had retained contact with the contaminated trousers that I couldn't take off. I think I skinned my butt when I slid forward during the crash, and the skinned area was burned right away, which is why it was hurting so much in those first few minutes and hours after the crash. The divers that recovered Todd's body said he was severely burned all over, and the medical examiners said that he had broken his neck upon impact.

The discovery of burns on my backside meant that I had to lay on my stomach most of the time, which was difficult with the two bulky casts on my legs. To say that I had an uncomfortable time of it for the next few days is a major understatement. It was also a little embarrassing to have my naked rear end sticking up in the air most of the time so that the proper medication could be applied. As an added attraction, in order to receive the required therapy for the burn, my rear end had to be dangled in a whirlpool for thirty to forty-five minutes each day. This was accomplished with chains and pulleys that allowed them to dunk my backside while my body was held in a v-shape as I rested the two casts on my legs on the edge of the whirlpool tub. After that was over with, the nurses would come around and peel the loose skin off my butt. Lots of fun.

They had a wonderful staff of nurses in that hospital, and they all maintained a good sense of humor about my predicament. One of the night nurses that JM and I referred to as "the Trooper," because of her tough demeanor and physical prowess, walked in early one day while one of the day nurses was still with me.

The day nurse said to the Trooper, "Have you met Miles?"

"No, we haven't been formally introduced as yet, but I know his ass well!" was the Trooper's quick reply.

I had two more operations during which Dr. Bendowski performed more magic on my extremities. He told me that the big hole in my left shin would fill in eventually, at which time I would be able to get a skin graft for it. The second operation went just fine, much like the first one. They had trouble getting the anesthetic to take for the third operation on 16 August, and I had some problems waking up, which caused a little panic on my part.

I was awakened one day by a phone call from a reporter for National Public Radio (NPR) who wanted to talk about the crash. He cited statistics that explained how our crash related to the other bad news chalked up by the aviation industry in Alaska that summer. We talked for quite a while, and it never occurred to me that he was recording our conversation, probably because I was half asleep. I repeated the story of the crash, which, by then, I had told many times.

I learned later that this NPR interview had been broadcast nationwide. JM received a call from a friend who had heard the program in Washington, D.C. The caller was one of our ex-employees. As you will see later, we have a lot of those running around loose.

"I just heard someone that claimed to be Miles talking on NPR about an airplane crash in Alaska, but I know it wasn't him, Miles is much more articulate than that!" That was her assessment of my first venture into nationwide radio broadcasting.

Listen, how articulate would you be if you had just woke up, were lying on your stomach with both legs in heavy casts propped on the bed frame, and with your naked behind sticking up in the air? I'm only superhuman, you know!

I was sleeping at midmorning on 19 August when I woke up to find an older couple standing silently by my bed. They were Todd's parents, paying a visit from Nebraska.

We exchanged greetings, and Mrs. Hatfield said, "You know, Dr. Hayes, they sent my boy's body home COD."

"Now momma," Mr. Hatfield said, as he hugged her quietly.

She wanted me to tell her about our last day with Todd, so I told her about breakfast at Margie's and a little bit about the flight. I said that I thought he did the best that he could under the circumstances.

"Yes, Todd was a very good pilot. If you knew Todd well, you would know that he would have wanted it to be him that died and not you," she assured me. And what do you say to that?

So the bean counters sent Todd's body home COD. They sent me bills for years afterward trying to charge me for the flight on the day the company almost killed me. That was one unpaid bill that I never worried about too much.

On 22 August, JM gathered our belongings for our trip home, which included a one-night layover in Seattle. These belongings included fifty-six get-well cards that she had taped all over the walls of the room, and about two hundred slides that had been under water for a week in the plane before they raised it. I had taught a field training course for British Petroleum (BP) on the south-central Alaskan coast at the end of July before coming to help JM with the mapping. I hadn't bothered to take those four lecture slide sets out of my briefcase nor take the briefcase out of the plane before we took off. JM put the film in plastic bags full of fresh water. After we got home, our graphics department remounted them, and I still use those slides. So if your slides ever stay in the ocean for seven days, take heart, they can be saved.

Next day, we went home to Columbia, South Carolina, where the company had made arrangements for me to stay in a Residence Inn located relatively near the center of town. We agreed that it would not be a good idea for me to go to our place out in the country, which we sometimes call "Deep River Bluff," or "the Bluff." Our house, which sits on the edge of a 250-foot bluff overlooking the flood plain of the Congaree River, was not exactly designed with invalids in mind. At the end of the month, JM had to go back to Alaska for another survey of the *Exxon Valdez* oil-spill site.

I had several visitors at the Residence Inn on Saturday, 1 September, including JM's mom and her husband, Bill Vining, and my two brothers, Edwin and Kenneth, and their wives, Nell and Betty. They all listened intently while I told them the story of the crash in detail. Bill seemed to be particularly enthralled, and he has referred to that story many times since.

"It's amazing what the will to survive can do," said my brother Edwin, and then he asked if they could do anything for me.

"I would like to go over to one of the grass fields on the university campus and try out my new fly rod," I was quick to suggest.

Under Danny Domeracki's guidance, the guys in the office had bought me a new nine-foot, four-weight Sage fly rod, one of the best money can buy (within reason), as a reward for surviving the crash.

My family drove me to the campus and stood around the wheelchair as I rigged the rod and made a few practice casts across the grass. The rod had a heavenly feel to it. Edwin made a few casts, and I noticed he was getting more distance than I was, but heck, I was sitting down. I had gone fishing with him a few times as a boy, but I was only seven years old when he went off to World War II. I received most of my fishing instructions from my Uncle Luster and Grampaw Worley, but that's another story. I was already dreaming about fishing again in Idaho and southern Colorado or on my beloved Davidson or Whitewater rivers up in North Carolina.

I spent much of the next three months in a wheelchair recovering from two more operations. But, as November turned to December, I was feeling pretty good, and JM and I were making plans for me to accompany the field team during the January 1991 field survey of the *Exxon Valdez* oil-spill site. Our field work was to be staged off the U.S. Coast Guard buoy tender the *Sweetbrier*.

We discussed the spill one morning on our way to work. Our initial report on the spill was nearly finished, and JM was being asked from time to time to comment on the damages caused by the spill. I wondered aloud how it would ever be possible to assess those damages.

"It made a pristine area no longer pristine," she said.

"It destroyed a way of life," I responded.

I was thinking about a meeting in Cordova, Alaska, in 1977 when I laid my head on the table and said nothing. But I am getting ahead of myself again.

On Saturday, 1 December, we worked all day at the office on the *Exxon Valdez* report that was due in a couple of weeks and returned to the Bluff about 7:30 P.M. After dinner, I was lying on the couch looking out through the glass wall and over the deck and the swamp, when the full moon broke out from behind a cloud and was shining directly through the glass into the house. We turned off the lights and watched the moon through our binoculars for a while.

Maybe it was the moon. JM and I lay together talking for a long time.

I went back over what had happened during the two hours that had followed the crash. It was one of the first times that I had tried to express myself about Todd's death and about those exacting few moments when I was trapped inside the plane.

"Everything was just right for you to make it," she said, "A little further offshore or a little further onshore and you would have drowned, or the plane would have burned."

I talked about what I was thinking as I crawled up the rock and during the two hours that I waited up on top. Then I said that I had tried to figure out why I had made it, and what significance, if any, there was to the fact that I was still here.

"I'll bet you have," she answered quietly, and we cuddled on the couch for a long time, enjoying the moon until it passed behind the edge of the rooftop.

A Primer on Oil Spills

I'm sure glad we're not shipping PCBs in million-barrel tankers.

Jacqui Michel (JM)

Tuesday, 19 November 1996—
Kuwait Plaza Hotel, Kuwait City

I was sitting in my hotel room thinking back over my day at the "International Conference on the Long-term Environmental Effects of the Gulf War," including the effects of the world's largest oil spill. The day before I gave my talk, reporting that the heavily oiled intertidal zone just north of Jubail, Saudi Arabia, hadn't changed much in the two years following the spill, which occurred in 1991. Our team's last survey found that in February 1993, fifty-two percent of the seventy-one subsurface oiled sediment samples that we collected were still relatively unweathered and, hence, still toxic to living organisms. Ours was not a biological study, but it was pretty clear that very little was living in that heavy oil. (This topic is discussed in more detail in the chapters "The Mother of All Spills" and "Ramadan," in Part 6.) Strangely enough, that particular area of the Saudi Arabian coast had been declared a wildlife refuge just in time for the spill. The refuge project, sponsored by the Commonwealth of European Communities (CEC), funded a group of European scientists to inventory the biota in the area. We chose that same area to do our survey because that was where a large percentage of the spilled oil went. A Brit named David Jones, working with the CEC program, gave a talk in which he acknowledged that the biota were affected by the oil with low counts being recorded during the first two years after the spill, but he stated that most areas had "recovered" by the time of their 1995 sampling period, four years after the spill. This prompted the chairperson of that session to say: "The Gulf was impacted less and is recovering more rapidly than people had predicted."

I was co-chairperson of that session, but I wasn't doing any talking about the spill. I was just trying to limit each speaker to his allotted twelve minutes—unsuccessfully, I must admit. So while the chairperson was able to make profound observations like the one I just quoted about the Gulf War spills, the co-chairperson only got to say things like, "Uh, I'm sorry Friedhelm, but you have only another thirty seconds. Please wrap it up quickly."

Despite my pique over the role I played in that session, the chairperson had a point—*the ultimate impact of any oil spill will never be as bad as that projected by the media.* I guess he was talking about the media. I don't know who else might have been making such predictions.

Another thing you can depend on—*we will never know exactly what that spill did to the ecosystem of the Arabian Gulf.* Why? Because the baseline data on the Gulf was so poor, and because the sampling plans and research carried out under the stress of the great catastrophe were so poorly funded and executed. Yep, we are still guessing. As with most big spills, the answer to the question of whether the world's largest oil spill did any harm to the Gulf is probably both *yes* and *no.*

Andrew Price, a veteran of many years of working in the Gulf, put it a better way in the leadoff speech in that same session: "The whole Gulf ecosystem hasn't completely crashed [because of the spill]."

True for sure. That assessment left Andrew room to speculate about what happened to the shrimp population and all the other things we don't have a clue about.

Listening to those talks, I wondered why scientists were looking all over the Gulf for impacts from the spill. We knew from satellite data and other information that the oil movement was restricted primarily to the Kuwait area and to about one-third of the way down the Saudi Arabian coast. Some people now think that some oil went as far down as Qatar. Not all over the Gulf! True, we first thought the oil might go all over the Gulf, but we knew it hadn't when the research cruises were finally carried out. Is there any wonder that no impacts were discovered in the areas where the oil didn't go?

A year earlier, I carried out a four-week tour around the United States giving a speech about what we had learned in twenty years of responding to oil spills. (This tour is discussed in more detail in the chapter "Fool on a Holiday"). During the tour, I was asked three questions over and over, and I provide my stock answers to those questions for this primer on oil spills to put you a little ahead of the game:

Q. Does the lack of conspicuous negative impacts to the biota around the natural oil seeps in the ocean near Santa Barbara, California, prove that oil spills are not a serious threat to the environment?

A. No. This is an old saw used frequently by industry apologists in the early days of the business. I guess thousands of years of evolution has allowed the animals and plants in the area to adjust to living with the oil. Also, when the seeped oil reaches the water surface, it is highly weathered. These seeps are not the same thing as dumping five thousand barrels of fresh crude oil on a tidal flat in unpolluted waters in France or southeast Alaska.

Q. Oil was spilled all along much of the U.S. shoreline during World War II from tankers sunk by German submarines, and now we see no trace of it. Doesn't that prove that oil spills do no long-lasting harm to the marine environment?

A. No. It seems that every old boy in the audiences during my lecture tour wanted to make this point. In fact, compared with the large tankers plying the oceans today, those tankers were quite small.[7] According to a study by MIT in 1977, of the fourteen tankers sunk off the Outer Banks of North Carolina during the war, the shoreline was hit by oil from only three of them.[8] The rest of the oil was carried out to sea by oceanic currents. Apparently, these spills weren't nearly as big as some of those experienced today.

Q. Did the *Exxon Valdez* spill significantly impact the ecosystem of Prince William Sound, Alaska?

A. Yes and no, for basically the same reason I gave for the Arabian Gulf. At a small scientific conference on Prince William Sound in Cordova, Alaska, that JM and I attended a little short of one year after the *Exxon Valdez* spill, biologists from the University of Alaska reported on data collected on the phytoplankton and other fauna and flora of the Sound over a several year period prior to the spill. These data showed that the biological populations in the Sound go through tremendous variability in their numbers on a year-to-year basis. Sorting out the impacts of the spill against such a background of change is no mean feat. Of course, some impacts were quite obvious, such as the hundreds of thousands of birds killed by the spill and the subsurface oil in the gravel beaches that was still sheening into the Sound five years after the spill (discussed in detail in the chapter "The Day the Music Died"). But if you flew over the Sound in a small plane today, you would be hard pressed indeed to note any vestiges of the oil spill. Therefore, the answer is yes and no, or we'll never know, or something like that.

JM responded for NOAA to a spill in January 1996 off the coast of Rhode Island from the barge *North Cape,* which released 828,000 gallons of home heating oil into the water column during a storm. Although this type of oil of all the petroleum products is the most toxic to marine organisms, scientists studying the spill were not able to detect population changes as a result of the spill for any organisms except lobsters. Once again, the problem of natural variability came into play. However, lobsters did have a high mortality, with twelve million juveniles perishing. Strangely, it does not appear to be only the oil that killed them. In part, they were narcotized by the oil to such an extent that they lost their ability to hold to the bottom and were eventually washed up on the beaches where they died.

And how about the following question that I heard Jane Pauley of NBC-TV ask an expert on national television. (I often wonder who these oil-spill experts are that you see on national TV and how they got to be experts since we don't ever see them at any oil spills.)

Q. Was the *Exxon Valdez* oil spill America's Chernobyl?
A. Maybe the stupidest question I've ever heard in my life, and I worked in academia for eighteen years. How can you compare a nuclear disaster of that magnitude with its associated human mortalities, medical injuries, pollution of crops, animal deaths, and so on with an oil spill?

I guess what I'm trying to say is that we still have a lot to learn in this field, and it seems like we learn something new at every major new spill. JM has a favorite expression, "I've never been to the same oil spill twice," which, I suppose, is the main reason why she still spends a bunch of weekends each year away from home tramping around in oil.

We can probably count on having many more oil spills in the foreseeable future. According to the 1996 *Oil Spill Intelligence Report* (OSIR),[9] thirty-one hundred oil spills of more than ten thousand gallons were reported for the interval 1978–1995. We know for sure that many spills that size have occurred in the Soviet Union and elsewhere that have not been reported. Our company, Research Planning, Inc. (RPI), has been responding to oil spills for NOAA since 1978. At the present time, we are called to help with the response to about fifty spills a year. We actually visit the spill sites in the field about five to ten times per year. In other words, we are usually pretty busy.

Six of the major oil spills we have studied, all of which are discussed in this book, are listed in the following table:

Table 1

SPILL	EVE*	SHORELINE IMPACTED (KM)
Metula (1974)	1.5	250
Urquiola (1976)	3.0	215
Amoco Cadiz (1978)	6.2	320
Ixtoc I (1979)	13.0	265 (U.S. only)
Exxon Valdez (1989)	1.0	2,100
Gulf War Spill (1991)	20.0	640+

* Exxon Valdez equivalent (EVE) = 10.8 million gallons.

A graph from the 1996 OSIR (see Figure 1) shows that the occurrence of spills is anything but consistent, with some years being relatively free of major spills. That same publication lists fifty-seven incidents between 1978 and 1992 that spilled more than 10 million gallons, with our very own "Chernobyl" spill, the *Exxon Valdez*, ranked number fifty-seven at 10.1 million gallons. However, the length of shoreline impacted was unusually large (Table 1). As fate would have it, our team at RPI has observed the oiling on the shoreline from the three largest oil spills on record: the Gulf War spill (1991, 240 million gallons), the Ixtoc I blowout in Mexico (1979, 140 million gallons), and the Nowruz field in the Arabian Gulf (1983, 80 million gallons).

The costs of cleaning up oil spills can be enormous. Etkin in a 1994 OSIR, "The Financial Costs of Oil Spills," stated: "A relatively small spill of less that 1,000 gallons can cost the spiller . . . millions of dollars if it occurs in an inopportune location, such as near a sensitive wetland area or a salmon fishery, or if the spill receives significant media attention."[10]

She also quoted an estimate of the average total value of cleanup, third-party damages, and natural resource damages, based on a study of historic spills. The average cost was $245.59 per gallon.

A spill of a little over 300,000 gallons of heavy fuel oil in the Tampa Bay area a few days before Labor Day in 1993 cost $40 million to clean up. At that spill, the primary concern was the impact on the tourist industry because fourteen miles of prime recreational beaches were oiled. The all-time record for costs was set by Exxon for the spill response and

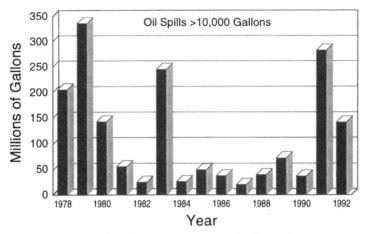

FIGURE 1: *Major oil spills (over ten thousand gallons) that occurred worldwide between 1978 and 1992.*

cleanup of the *Exxon Valdez* spill, a total of $2.1 billion. That number doesn't include the third-party damages and other legal costs.

In addition to the financial costs, serious human costs are sometimes associated with the accident that caused the spill and with the spill response itself. Jerry Galt and Gary Shigenaka of NOAA HAZMAT (the Hazardous Materials Response and Assessment Division of NOAA) recently returned (early February 1997) from a site visit at the request of the Japanese government to the *Nakhodka* spill of 1.5 million gallons of light oil in the Sea of Japan. In their trip report, they stated: "Because of the deaths from cardiac arrest of four volunteers and the approximately 230 injuries related to dermal exposure, eye irritation, and slips and spills on the shoreline, the Japanese are very concerned about the human health effects of the spilled oil."[11]

Analyzing an oil spill always requires an in-depth knowledge of its chemistry. JM has a minor in geochemistry, one of the reasons she is so talented at responding to spills. The types of oil spilled varies from crude oils, which can range from lighter strains (for example, Kuwait crude) to heavier strains (some of the Venezuelan crudes), to a host of refined products, such as diesel and Bunker C (a heavy product used to fuel many oceangoing vessels). Almost all crude oils *float*, which certainly helps in the containment and recovery business. However, these days petroleum products that do not necessarily float when spilled are on the market. Two of the more recent spills RPI has responded to, the *Morris J. Berman* spill in Puerto Rico in 1994 and the Tampa Bay spill of 1993, were of heavy oils, some of which sank to the bottom, taxing the creativity of the cleanup crews.

A few of the more important chemical considerations about spilled oil include:

- Lighter fractions of the petroleum compounds are the most toxic.

- Lighter fractions may be lost quickly by evaporation.

- Chronic toxicity is reduced by microbial degradation of the oil.

- Penetration and burial of the spilled oil in sediments slows the weathering process by an order of magnitude.

These facts mean that diesel and light fuel oils are the most toxic oils spilled; however, they evaporate quickly and long-term cleanup problems are rarely encountered with spills of these products. In addition, the longer the oil weathers, hence losing the lighter fractions, the less toxic the oil becomes. Finally, one of the more difficult problems to deal with occurs when the oil penetrates or is buried deeply in the sediments of the intertidal zone, a situation encountered at both the *Exxon Valdez* and Gulf War oil spills.

Back in Kuwait, I had begun the second day of the conference with one of my ex-graduate students, Dr. Mohammad Al-Sarawi, who had just recently been named to head Kuwait's environmental protection agency. We played hooky from the meeting so we could take a tour of the Kuwait City waterfront. The twenty or more kilometers of shoreline had been completely reconstructed under a master plan created by a team of U.S. architects and scientists, of which yours truly was a member, in a project which had started almost exactly twenty years earlier to the day. From the top of the Kuwait Towers, we could look down on the completed structures of "the hook," Green Island, and the artificial beaches, which looked just as architects Frank James and Paul Pawlowski had drawn them back in 1977.

Back in my room at the end of the day, I read the conference conclusions, which were published under the headline "Research Gaps Worry Experts" (don't they always) on the front page of the *Arab Times*. The major headline for that edition of the newspaper read "Allies Manoeuvre A Message—Repel Mock Attack." You probably recall that the U.S. Army was back in town in response to Saddam's latest test of the will of "the Allies". Beside the headline was a picture of the youngish-looking American ambassador to Kuwait, Ryan Crocker, walking along with the troops in the exercise area. He had the same blank expression on his face in the photograph that he had retained throughout the reception for our conference held at the Kuwait Towers the night before.

When I finished reading the conclusions, I thought back to the last session of the conference, which had taken place about three hours earlier. A few minutes before the end of the session, the hooded face of one of the Kuwaiti organizers of the meeting appeared over my right shoulder and said: "Dr. Hayes, would you like to make the final speech of the conference as the spokesman for all the participants who came here?"

"Oh no. No. No. No, no, no. In a word, no."

He went away. I was not going to volunteer again, not after that rotten duty as timekeeper.

Nonetheless, I had been thinking about raising my hand and making a point that was bothering me, but I couldn't quite think how to say it right without sounding like an arrogant buffoon. I gave up and went through a little fantasy speech in my own mind that went something like this:

Mr. Chairman, my name is Dr. Space Wanderer. I have dropped by your conference on a short visit from the planet PAH (sometimes known as polynuclear aromatic hydrocarbons), which is located in the galaxy Dibenzothiophene. It is my job to report back to my superiors on your progress in determining the impact of oil spills on your fragile planet, especially here in the Gulf, which is quite similar to some of the water bodies on PAH. This is not the first conference of this type that I have attended, so I am not very surprised that you still don't have a clue what the spill did to the Gulf. How could you? You had no clue about the functioning of the ecosystem of the Gulf before the spill. However, you shouldn't feel badly about this; similar ignorance exists among the scientists that inhabit other planets throughout the universe. I have to admit that, because of intergalactic travel lag, I did not attend the lectures this morning because I have been finding it very difficult to stay awake through some of them. Instead, I took a quick overflight of the Gulf, using technology that I am not at liberty to disclose to you, all the way from the Shatt al Arab (Tigrus/Euphrates delta) to the Strait of Hormuz. On the basis of these observations, I have a question for you. Why have you declared war on the intertidal zone?

By making this point, I do not refer to the oil spill, which, of course, didn't do the intertidal zone a whole lot of good. I don't have the exact numbers, but I am sure that many more hectares of intertidal habitats along the west shore of the Gulf have been buried by rubble in land "reclamation" projects than were impacted by the spill. In fact, as you well know, the whole twenty kilometer waterfront of Kuwait City is built on rubble from the old city that was pushed out on the intertidal zone before construction of the modern city began. The same thing can be said for much of the outer shoreline of the United Arab Emerites and the southeastern shore of Saudi Arabia. And the process continues at ever increasing rates. One would have to conclude that the intertidal zone, the feeding areas for migrating birds and the breeding grounds for a number of marine organisms, has achieved zero value in the minds of those plotting the future development of the coastal areas of the Gulf countries. Not so?

Of course, I didn't make this point out loud before the conference. In fact, I didn't ask one question the entire time. Why? Because I'm a pretty shy person who has never been known to spend a lot of time in this activity.

At the end, one Dr. K. H. Batanouny from Egypt, a lover of dune plants (we talked about it on the bus ride to the meeting), made a bombastic closing statement on behalf of me and the other participants, expressing our appreciation to the organizers of the meeting for their hard work and excellent organizational skills.

"Wish I could do that," I thought.

Once more, back in the Kuwait Plaza Hotel, I decided to relax a while in anticipation of my 7:00 A.M. departure for the airport the next morning and the twenty-six-hour trip back to Columbia, South Carolina. I wanted to finish a book I had been reading, *The Sunfishes*, by Jack Ellis.[12] This book is not scientific; it's a fishing book. The final chapter started in a pretty heavy vein so I stopped to watch TV for a while. TNT was showing *Andy Hardy Goes to College*, or something like that, starring Mickey Rooney. A swell movie, by golly. Those were rosy times—they weren't thinking about global warming or oil spills back then. About fifteen minutes of that light-hearted fare was enough for me, so I returned to Jack Ellis's book, starting at the point where he was bemoaning the fact that wild fish are disappearing from the face of the earth, and that one would have to live in a vacuum to be optimistic about the future of wild fish. He also intimated that man's own existence is in some doubt, noting that journalist Linda Ellerbee commented some years ago that 1955 was "the last year we could look into the future and not be afraid of what we saw."

In 1955, I was playing second base for the Berea College Mountaineers, and Mickey Rooney had moved on to much less optimistic roles in the movies. But right then, 20 November 1996, I was reading a book trying to learn how to tie some killer bluegill flies. My ex-student, Mohammad, was heading in his Mercedes to his beach house for the weekend so he could take his "small boat" offshore the next day and fish—for wild fish, presumably—with his wife and catch a fine dinner. Earlier that morning, Mohammad had shown me the futuristic buildings and monuments of the new Kuwait rebuilt after the Iraqis left. I didn't see any starving people lying in the streets, but I saw lots of marble buildings and architectural masterpieces. We had piles of food from all over the world at the luncheon smorgasbord at the conference. Furthermore, the impacts from the Gulf War on the environment of the Arabian Gulf "were less than had been predicted."

So what's your problem, Jack?

PART 2

CHILD OF THE

KING

My Father is rich in houses and lands,
He holdeth the wealth of the world in His hand!
Of rubies and diamonds, of silver and gold,
His coffers are full, He has riches untold.
I'm a child of the King, A child of the King:
With Jesus my Savior I'm a child of the King.

Sung by Old Man Woods on a summer evening in 1951
at Brother Dycus's weekly service at the Biltmore
Estate Dairy, Asheville, North Carolina [13]

Oconaluftee

Saturday, 20 October 1934—
Oakley, North Carolina

I was born and grew up in a four-room house located below a dirt road, called School Road, in the community of Oakley on the outskirts of Asheville, North Carolina, birthplace of Thomas Wolfe and Charlie "Choo Choo" Justice. The house was reconstructed in 1932 by my father, Norman Eustace Hayes, a graduate *summa cum laude* from the fifth grade of a small country school located a little farther back in the mountains (in Hayes Cove, I believe).

As far back as I can remember, I have been a tree-hugger. My father, who was a superb gardener, knew the names and songs of all the birds in our county, collected rocks, hunted bee trees on his Sunday afternoon walks in the mountains, and pondered the meanings of all things natural. He was a gentle man, and about the only time I ever saw him get mad was when he started complaining about the *dad-burned* polluters who were ruining all the streams. It was the "*dad-burned* tanning company and the *dad-burned* paper mill that were turning the Pigeon, Tuckaseegee, and the French Broad rivers black and killing every living thing in them," he said. And when he started talking about the *dad-burned* politicians who were bought off by the *dad-burned* foreign mill owners so they could keep killing the rivers, he became extremely upset.

When I was in the fifth grade, my mother gave me a book for Christmas, *Birds of the South*,[14] and pretty soon I could keep up with my father in the bird ID business. But my first love in the natural realm was fishing, and I had the best of teachers, starting virtually as soon as I could walk. First were Grampaw Worley and Uncle Luster, my mother's father and brother, for whom I dug worms for the afternoon's fishing. As soon as Luster got home from the mill, we headed for some place like the Rocky Broad River, where Grampaw almost invariably caught three or four huge redhorse suckers on the worms and Uncle Luster demonstrated the fine art of fishing with flies for trout using a classic bamboo fly rod, his prized possession.

As great a fisherman as my Uncle Luster was, he had two serious flaws as a piscatory guru: an awesome temper and almost no patience. If the fishing was good, everything was fine. However, if it was slow or he broke off his fly on the biggest fish of the day or he hung up in a tree or something of that sort, he went into a minor rage and said things like, "I'm going to throw this rod and all this gear right out into the middle of the river and never go fishing again as long as I live, so help me God!"

On several of those occasions, I ran over, grabbed his tackle box, and said, "Don't throw it Uncle Luster. I would really like to have this stuff!" At which, he usually calmed down and started fishing even harder than ever.

Before long, I built my own fly rod out of one of Grampaw's cane poles and was stepping out with friends like Marvin Jones, the local Huck Finn who dropped out of school early to become one of the most adroit bait fishermen I ever knew. Then there was the time when I was about eleven or twelve years old that I walked to the Greenwood Hole on the Swannanoa River with another mill hand, Hoojer Smith, and he caught a whopper smallmouth bass that I proudly lugged all the way home draped over my shoulder. Hoojer wanted us to detour by Hamrick Feed and Seed where we could show off the monster he had caught on the live nymph of a dobson fly (which we called a "grampus," also known as a "hellgrammite"). When the hangers-on who always gathered at the feedstore asked Hoojer what he caught it on, he told them "a jitterbug," which is a big artificial bass plug. My father and I laughed and laughed for weeks about Hoojer's little indiscretion.

Soon after that, I became friends with a school mate, Adrian House, and we carried our homemade fly rods all over the hills, fishing whenever and wherever we could. I read all the fishing books in the Buncombe County Library, and somebody gave me Ray Bergman's book, Trout,[15] for Christmas, from which I learned considerable about fly fishing, spending hours looking at the colored plates of the hundreds of flies pictured in that classic. Before long, following the admonition of Lee Wulff in one of the fishing magazines that Adrian and I hoarded, I began practicing "catch and release." Amazing as it was for a young country boy to "throw back" all the fish he caught, I believe that practice highly pleased my father, who was not much of a fisherman himself.

I was named after Granny Worley's father, Miles Goforth, a mountain man who fought in the Civil War. Granny always insisted that he fought on the "winning side." I laughed at that suggestion because I knew that the South lost the war, which was about all I learned about that altercation at the Oakley School. It was not until about twelve years ago, when

JM and I became Civil War buffs, that I learned that a large contingency of the mountain people who didn't own slaves were loyal to the Union. In fact, a massacre of some of the loyalists by the rebel army occurred not too far from where Granny Worley grew up, Big Pine Creek. I do not want to know more about my great grandfather's political affiliation. I prefer to believe he was a Lincoln Republican who fought on the "right side." Why? Because Abraham Lincoln is my number one hero—well, okay, after my father, my brother Edwin, and my cousin Glen. My father didn't think much of my namesake, referring to him as a "shiftless jack-leg preacher."

Music and religion were also big in my family. My mom and my two sisters, Judy and Norma, could all play the piano, and my father once sang tenor in a religious quartet. That quartet even made a record, which I still have in my possession. My brother Edwin formed a country music band when he was in high school, but World War II interrupted any ambitions he might have had in that direction. My Uncle Junior, my father's younger brother, toured the country with a hillbilly music band for a few years. We often gathered around the piano and sang hymns in the evening, and, if I do say so myself, we were pretty good. All of us, at one time or the other, performed songs in our local church, which was Southern Baptist.

In the spring of my fourteenth year, I believe it was, I arrived home after dark following baseball practice and walked through the kitchen, which was abandoned, to the living room, which was lit by a single naked light bulb that hung from the center of the ceiling. Standing on a small stool with her head almost touching the light bulb, which cast a halo of light around her long brunette hair, was a beautiful thin girl dressed in a long white dress that my mother and sister Norma were measuring. I was struck dumb and instantly retreated out of the light. Then I watched silently for a minute or two before escaping to my back room.

The next morning, I asked Norma who the girl under the light bulb was, and she told me it was Raye Dycus who, along with her twin sister Faye, had been taking piano lessons from Norma for several weeks. Noting my interest, she remarked that if I wasn't always off somewhere playing ball or fishing, I might get to meet her.

Well, I learned their schedule and made sure that I was home the next time, and we met and eventually became sweethearts, a first for both of us. And Adrian met Faye and eventually they became sweethearts, too. Raye and Faye quickly learned to play the piano, and they could both sing. After a while Norma came up with the idea that they, my sister Judy, and yours truly singing bass, would make a good quartet. We started practicing together and before long were making appearances in several of the local churches. We even appeared on a couple of Sunday morning radio programs.

Raye's dad was a preacher who had a unique and admirable philosophy, I thought. He entered the ministry but didn't quit his day job, which was painting houses. He said that if somebody was paying his salary, he would be obligated to say things in his sermons that the people paying him wished to hear. He preferred not to do that; he wanted to say what the Lord, not his benefactors, told him to.

His day job was painting the numerous buildings and facilities of the Biltmore Dairy, which was located on the famous Vanderbilt Estate, the place where Grampaw Worley learned horticulture while working for Vanderbilt's heirs. Brother Dycus witnessed on the job, and some of the other employees of the dairy suggested that he hold evening services for them. He did, and soon he had a quartet that performed free every Tuesday night. Brother Dycus's favorite number was a religious rap rendition of Moses leading the Israelites out of Egypt, with me doing the talking while the three girls hummed in the background. That number was a big hit everywhere we sang it. We weren't the only ones that sang at Brother Dycus's services, however. Anyone who got the spirit could walk right up front, testify, and sing. Old Man Woods performed frequently, giving touching versions of "Child of the King" and "Lord Build Me a Cabin in the Corner of Glory Land."

During those halcyon days with Raye, I spent many nights with my Grampaw and Granny Worley in order to help with the chores—milk the two cows, feed the chickens, and so on. Grampaw was considerably slowed by a stroke. During the long, cold winter nights, we sat before the fireplace and popped corn. Grampaw often told stories of his early days on Big Pine Creek, far back in the mountains where his ancestors had settled two generations earlier. He told ghost stories and stories about the wild animals along the creek. I never tired of those stories. Big Pine will always be a special place for me.

Big Pine Creek had been settled by the whites for a long time, and it wasn't even a wild trout stream any more when I came along. About thirty miles from there were the clear waters of the Oconaluftee River that originated in the high mountains of the Great Smoky Mountain National Park and then ran through the Cherokee Indian Reservation.

Brother Dycus, in addition to having a unique philosophy about paid liars, also had, for the place and times, a novel approach to race relations. He frequently changed positions with a black preacher who could really bring down the house, and he also fraternized with the Indian preachers on the reservation. Naturally, then, our quartet performed on more than one occasion on the reservation. I particularly remember one time when we were involved in an "all day singing and dinner on the grounds," with the Indian quartets singing in the Cherokees' native language and us in hillbilly English. The church we sang in was located alongside the Oconaluftee River, so I took every available opportunity to walk on the nearby swinging bridge and look for trout. I had never seen water so clear. It had the transparency of air. Pools over your head looked two feet deep! The Oconaluftee struck me as being the holiest of waters, waters that the Spirit Father had created just for the maligned Indians, most of whom lived in one-room wooden shacks.

My image of the Oconaluftee River from that swinging bridge will always be the measuring stick by which I determine the purity of nature. A few streams in the High Sierra and in northern Idaho measure up, but not many. The beaches on the outer coast of the Kenai Peninsula of south-central Alaska do, too. In the Oconaluftee, I had found perfection, a counterbalance to the polluted black rivers that bothered my father so much. As a fourteen-year-old working in the sawmills making boards from the virgin timber being stripped from the mountainsides, he ventured into the mountains long before me. No doubt he had his own Oconaluftee, which is why he so fervently hated what was being done to our rivers.

I still go back to the Oconaluftee River to fish now and then. Nowadays, almost the entire riverbank inside the reservation is cluttered with tourist shops and motels, and the Indians stock the river several times a week with pond-raised, fish-meal-fed behemoths for the tourists to catch with cheese balls. JM and I go high into the park and fish Raven Fork of the Oconaluftee, which still has clean water that even contains some native brook trout. But the fish are small, the insects are sparse, and the pH is dangerously low due to acid rain.

At Berea College, I started to major in agriculture, wasting a couple of semesters taking courses such as agronomy and animal husbandry. I thought that I wanted to be a farmer, but I soon learned that I couldn't do that because I wasn't rich enough to buy a farm. The thought of being a county farm agent didn't appeal to me in the least. A course in physical science, which included several segments of physical geology, revealed the promise of what I was looking for—a chance to work outside, a chance to see the world, and a chance to get a Ph.D. in science. I've never regretted for one second my decision to major in geology.

The Christmas vacation of my freshman year I went home, and Raye and I broke up—not because we weren't still in love, but because we had a timing problem. She still had two years to go in high school, and I still had about at least eight more years to go in college. I couldn't see that far into the future.

When I went home for Christmas my junior year, I visited Granny Worley, who was then living with Uncle Luster. Grampaw died about a year before I left for college. She had been attending some of Brother Dycus's services at which Mrs. Dycus gave her something to pass on to me. It was my high school graduation ring that I had given to Raye.

That night I drove around for a long time in my father's '51 blue Plymouth, stopping by the school to walk across the ball field and along the roads near home, thinking about Raye. Then I drove from the school down the hill past the mill and crossed the bridge spanning the Swannanoa River. From the middle of the bridge, I tossed the ring over the edge, and it dropped into the depths of a long pool located several hundred yards downstream from where Hoojer Smith caught the big bass. I was then completely on my own.

I stayed at Berea College for an extra year, partly because of the change in majors but mostly because I had the opportunity to be an assistant coach for the baseball team in my fifth year. At the local Baptist church, where I was director of the student union choir, I met Barbara Anne Hyder, who was two years behind me in school. We started dating in the fall of my last year, and at the end of the following summer, during which I worked as a groundwater geologist for the U.S. Geological Survey in Louisville, Kentucky, I went back to Berea and proposed marriage. During the following year, we made several trips back and forth between Berea and St. Louis, where I attended Washington University on a graduate teaching assistantship. We were married in June 1958, and she joined me in St. Louis and began training as a medical technologist.

In St. Louis, I learned to teach under a master craftsman, James C. Brice, who was in charge of the graduate teaching assistants. He also taught me sedimentology and geomorphology and supervised my master's thesis. Then I went to the University of Texas, where I fell under the inspirational spell of the one and only Robert L. Folk, now recognized as one of the greatest sedimentary geologists who ever lived. You no doubt heard about the recent discovery of the record of possible bacterial life in the rocks of Mars. He discovered a similar record of life here on earth in some rocks in Italy. Christmas 1996 we exchanged year-end messages. He scribbled a note saying I was finally right about something, Alaska was beautiful. He and his wife, Marge, made their first trip to southern Alaska the previous summer. In the main body of the letter, he referred to his discovery of the dwarf bacteria in Italy, and hence in the Mars rocks, as "dumb luck."

My academic record at Washington University was anything but outstanding, and I received only "provisional" admission at the University of Texas (UT). I went there anyway because Brice at Washington University told me I should work under Folk. Right away, Folk got me a job at the Defense Research Laboratory at UT, which allowed me to examine classified data on the shorelines and inner continental shelves of the whole world, probably the single most valuable learning experience of my scientific career.

For the next five years, I did nothing but creative work. It was the only time that I have had the freedom to pursue any new idea that came along. During the late nights when I was working with the other grad students and some of the younger faculty members, namely Al Scott and Earle McBride, we wrote papers, published field trip guidebooks, and came up with original ideas that hopefully have stood the test of time.

With the bulk of my dissertation work behind me, in August 1964, Barbara Anne and I put all of our meager belongings into a U-Haul trailer and headed for Amherst, Massachusetts, to a teaching job at the University of Massachusetts (UMass), where I had never visited. The chairman, H. T. U. Smith, had hired me sight unseen, based on the recommendation of my office mate Murray Felsher, a graduate of UMass, and some of the faculty at UT, I assume. The job was perfect, with the whole New England coast lying untouched before me, in a geomorphological sense, and a host of eager students chomping at the bit to get into this new field of coastal geomorphology. H. T. U. also put me on his grant for Antarctica research, and I headed down there for a six-week stint in December 1965. Being "on the ice" was a learning experience, but toward the end of the long stay, I came down with a urinary infection so severe that I lay in a U.S. Navy hospital bed in Christchurch, New Zealand, for several days, thinking that I would surely die any minute.

"Suzanne's Lament"

Wednesday, 9 February 1966—
Amherst, Massachusetts

My eldest daughter, Joy Elizabeth, was born on this date, which was only about three or four weeks after my sojourn "on the ice." My second daughter, Mya Suzanne, was born two and a half years later on 8 July 1968. During those early years, we rented a small research cabin on the beach at Plum Island, Massachusetts, where we spent most of our summers. The girls loved the cold water and the coarse sand of the steep beach.

By the time Mya was one year old, my graduate students at UMass and I were in the fifth year of research on the New England coast, and a pile of work was being produced, resulting in a regional field trip led by our illustrious team. We also started the Alaska work in the summer of 1969, but I had an operation that summer related to the urinary problem that I had developed in Antarctica, so I didn't spend that much time in Alaska.

However, in February 1970, Barry Timson, one of my graduate students, and I went back to Alaska for a few days to resurvey some of the permanent beach-profile stations that had been established along the ocean front of the Copper River Delta the previous summer. Working out of the small fishing village of Cordova, we hired a native pilot, Jim Foote, who landed a Cessna 180 on the beaches by our survey stakes. During this stay, we had a stretch of outstandingly clear weather, and I was able to take an extraordinary number of excellent photographs.

Partly based on my suggestion, the geology department at UMass had established a weekly seminar at which faculty and graduate students presented results of their research. A while after Barry and I completed the Alaska survey, I prepared a talk on the survey to present in the seminar. Wanting to try something different, or special, I put together a sequence of slides set to the words of the song "Let It Be" by the Beatles. The slides were set up in couplets: one slide of the pristine shore of Alaska we had just surveyed, and the other of some badly modified shoreline of the "lower forty-eight," emphasizing New England, New York, and New Jersey. I did my best to match the morphological/environmental setting in each couplet, and the contrast was striking. It seemed like a simple way to say let's preserve this little piece of the world that is still much the way the Spirit Father made it.

In addition to being basically shy, I had never been prone to expound my real feelings about much of anything in public. I thought I was a pretty good teacher, and I knew I could deliver a talk. I had even sung before audiences when I was younger, without showing much nervousness, but those performances were based on topics that I felt I was master of or took place after several rehearsals. What I was about to do was expose my innermost feelings before my peers and students for the first time by speaking out about something of importance to me that didn't necessarily have anything to do with geology. Why? That part of Alaska is still to this day the most special place on earth to me. I think the simplest way to put it is to say that during that spectacular field survey in February 1970, the Alaska coast made my heart sing.

I gave my talk, which wasn't exceptional, and then I played "Let It Be" and showed the slides. At the end of the song and slide presentation, the place was deathly quiet. I turned the lights on, and everyone began to file out, still very quietly. I was standing by the door. My friend and racquetball partner, Leo Hall, a Harvard graduate, was one of the first to reach me. He looked at me but didn't say a word. He just reached over and touched me on the shoulder and then went out. Several others did the same thing.

"My God, what did I do?" I thought.

Later in the spring, I went on a lecture tour of several universities and colleges along the east coast sponsored by the American Geological Institute. Where appropriate, I presented the "Let It Be" sequence after the lectures with much the same reaction I had gotten at UMass. Of course, it was different for me with those audiences because they weren't my closest associates. Nonetheless, I thought this technique should be explored further.

In the summer of 1970, our coastal geology group carried out a major field excursion back to the south-central Alaska coast. We had one team, led by Joe Hartshorn and Jon Boothroyd, studying the glacial streams, and I led the other team, which was responsible for surveying the hundreds of miles of shoreline between Cordova and what is now Glacier Bay National Park. I was assisted by four budding geology students— Paul Hague, a "mature" student and electronics expert who had served a stint in the U.S. Navy, Bob Henry, a UMass gymnast who had family connections with the Mafia in Boston (obviously on his mother's side), Frank J. Raffaldi, Jr., a wrestler and musician with no Mafia connections, that I knew about anyway, and Woody Hobbs, who was just a plain old youthful geology major at that time.

I chartered a plane and pilot for the summer from the Chitina Air Service in Cordova, which was founded by the famous bush pilot "Mudhole" Smith. "Mudhole" wasn't flying clients anymore, but we flew with his son a couple of times. The pilot assigned to us was Pat James, an ex-cranberry farmer from Minnesota who flew his first plane at the age of sixteen. Pat was an ace pilot who could land the tail-dragging Cessna 180 on almost any of the beaches we were surveying. He had a reasonable, cautious approach to flying, which didn't bother me much.

The six of us worked the low tides, often two a day because of the long hours of daylight. We lived in mobile homes, abandoned mining towns, and forest service cabins as we made our way south along the coast. The isolation of those living quarters encouraged us to become a close-knit group, and we talked philosophically about almost any topic you could think of.

I remember one of my first chores was to convince Bob Henry that he shouldn't be shooting the seabirds just for the fun of it. Later I tried to convince him that he shouldn't get drunk in the bars of Cordova and stand up and yell at the top of his voice, "I can whip any fishermen in this GD place!" which is something he probably could have done, considering his prodigious build, but we were sorely outnumbered, to say the least. Frank J. Raffaldi, being a wrestler himself, was also well built, looking like a slightly smaller Sylvester Stallone. They both called me "boss," which pleased me a great deal. Back at UMass, we would sometimes walk down the hall, Bob on one side of me and Frank on the other, and the path ahead of us would clear with alacrity as we approached. What a field crew!

Once the logistics of our survey work were finally worked out, I had another problem—namely, how on earth to study such a large piece of landscape in just a few weeks. The plans I made back home didn't seem to work in the field. Then one day while sitting in the wooden outhouse in the abandoned oil town of Katalla, an inspiration struck, which later became known as the "zonal method." This idea didn't win any prizes, but it did get us through the summer. Ed Owens, a graduate student, talked me into writing a paper on the subject (in fact, he may have written most of it) and presenting the idea at the Third Annual Geomorphology Symposium in 1972. We added the work of two other students,

Dennis Hubbard and Ralph Abele, to the paper, and it was the lead-off presentation in the symposium. The paper was entitled "The Investigation of Form and Process in the Coastal Zone." Some choice words from the abstract of that justifiably unheralded piece of work follow:

At the reconnaissance scale, a method of study (*zonal method*) has been developed which is based upon a systematic sampling program and upon the quantitative investigation of representative geomorphic units or zones. This approach has provided a rapid method for synthesizing critical environmental properties of large coastal areas, such as the glacial outwash plain shoreline of southeastern Alaska.[16]

I present this excerpt here so you will see that we did more than philosophize and take pictures during that summer, which was an evaluation that certain members of the heralded U.S. Geological Survey made of our work because we were so slow to publish the results. Or was it because they didn't like my new beard?

I must confess, however, that the words of "Let It Be" ran through my mind more than once, and I was certainly taking more pictures than were required by the contract. I was thinking about a slide show to go with the "Let It Be" sequence.

Some people said I "went crazy" in Alaska that summer, but it was a little more complicated than that. In Alaska I learned something new about life. I learned to feel. I felt the wind, and I saw what the Indians called the "Spirit Father" or the "Great Spirit." The natural things, the earth god—pure, simple, cold, wet, real. I loved the god of nature that I had rediscovered there, because it brought back the child in me. And in so doing, I was able to take pictures superior to any I had taken before, and to decipher nature apart from pure science.

That was the upside. On the downside, it is painful to even think about, much less write about, my return to Massachusetts in the fall of 1970. To make a long story short, my marriage broke up, and change was the order of the day.

Through all the chaos and heartbreak of that fall, I worked with Paul Hague and his wife, Chris, to put together an expanded slide show, or "tone poem" as it would later be called, based on the summer's work. I had a general theme in mind and provided most of the slides, and Paul was the chief musical consultant, providing much of the music and equipment.

At the end of one long session in which we sorted through piles of my slides and ones from the other field crew members and worked on the central theme, I heard Chris say about me, "I think the man loves the earth!"

Once the show was put together, we named it "Suzanne's Lament" and started making the rounds. At the beginning of each presentation, I said something like, "This show is in three parts. Part One is the introduction to the beauty and pristine character of an unspoiled shoreline. Part Two shows our field team reacting with that environment, and Part Three is an appeal for its preservation."

After the show was over, I would ask the audience if there were any questions. And the audience would invariably laugh—nervously. Usually there weren't any questions. Nobody in those audiences was looking for microphone time.

The name "Suzanne's Lament" was my idea. A favorite song of mine, "The Cowboy's Lament," was one of the few I could play on the guitar, and my daughters liked it as a bedtime tune. In case you don't remember that song, it's about a young cowboy who died of a gunshot wound on the streets of Laredo. One of my daughters was named "Suzanne," so you can figure that part out for yourself. The name "Suzanne" was mentioned frequently in many of the songs we used; James Taylor's "Fire and Rain" was one. In addition, the song "Suzanne," by Leonard Cohen, was a pivotal part of the story. Using mostly pictures of Kayak Island, which is the centerpiece of the unspoiled landscape in that area, we matched the words of Cohen's song to the land forms: "Suzanne took my hand and she led me by the river," and so on. In this way, "Suzanne" became the embodiment of Mother Earth.

There was a lot more to this show than an environmental statement, in my mind anyway. I showed pictures of the loved ones back home, including my children, in the slide sequence set to the words of "Golden Slumbers" by the Beatles. The show ended with Dave Van Ronk's version of "Both Sides Now" (which was Hague's suggestion), and the last words heard were "I really don't know life at all," which certainly summarized the confusion going on in my life at that time.

Why "Suzanne's Lament" was such a great success has always been a mystery to me. Something I read recently in Daniel Quinn's book, *The Story of B*,[17] may shed some light on this puzzle: "In 1950 this was something everyone could count on: Exploiting the world was our God-given right. The world was created for us to exploit. Exploiting the world actually improved it." When we presented our slide show in the early 1970s, people were beginning to question that concept.

Our first formal presentation of "Suzanne's Lament" was before the

UMass geology department on 16 December 1970. Two days later, we gave a show for the elementary school in Hatfield, Massachusetts, where Chris Hague taught. The 150 kids that turned out for that program seemed to embrace it. They wrote notes to Paul and me, saying they "loved seeing pictures of the river that looked like a snake" and things like that.

In the first three months of 1971, we made fourteen more presentations in the western Massachusetts area, including one before a packed house at Amherst College, where we spilled all the slides out of two slide trays onto the floor only ten minutes before show time. Luckily, I got the slides back into the trays only a couple of minutes late. Although we had a shaky start, when the program ended, we were treated to a standing ovation. With that behind us, I thought it was time to take the show on the road, so to speak.

I sent a note to a friend who was organizing the upcoming Northeastern Regional Meeting of the Geological Society of America to be held in Storrs, Connecticut, in March. I suggested that we could present our slide show at the meeting. He concurred, but, when we arrived in Storrs, I was disappointed to find that the only advertisement for the show was a hand-lettered eight-by-eleven-inch poster stuck on a few bulletin boards here and there. The posters stated that a lunchtime presentation of a slide show on Alaska would take place in a room on the second floor of a nearby motel on 18 March. We set the show up anyway, although the room, a standard motel room with the furniture removed, was so small that it couldn't possibly seat more than twenty people. The equipment itself, which consisted of two slide projectors, a dissolve unit, and a rather large and cumbersome sound system with two speakers, took up a lot of the space. Come show time, the room filled up quickly, and people spilled out into the hall, well beyond where they could see the slides.

After the show was over, my friend said that he would pass the word that "Suzanne's Lament" would be presented again at 5:00 P.M. in the convention's main auditorium. Word must have spread quickly because by five o'clock at least 350 people, most of the attendees at the convention, were waiting to see that presentation, which received an extraordinary reaction from almost everybody there. I did hear one crusty old Yale professor, author of a couple of major textbooks, say, "It would have been better if they had set the slides to Brahms." But this was 1971 and 60s music seemed a better expression of the mood of the times than Brahms.

I had never seen anything like the way people responded to the second performance. Professionals I had known for a long time stopped to say "Miles, thank you for that" and so on. Another show was scheduled for the next day at the end of the meeting, which was also well attended, with many people seeing it for a second or third time.

Unfortunately, because of the chaos of the times, I was not taking detailed diary notes. The following is the way I remember a conversation that Paul Hague and I had on the trip from Storrs back to Amherst on the afternoon after the last presentation. Paul was driving and I was looking out the window at the passing, neatly kept New England landscape, and I thought to myself, "If I don't ever do anything else in my life, at least I did 'Suzanne' and we made them all cry."

For a long time we drove silently, until finally Paul spoke up, "That was really something back there in Storrs, I mean the way they just kept coming back, with the audience for each show getting bigger and bigger. And I was really surprised by the way they were so quiet at the end."

After another long lapse into our own thoughts, Paul starting talking again. "I've been meaning to ask you something. You know at the end of that first time in the motel room. The show was over, but nothing was happening, and you had to get your friend, the chairman, to say something like 'maybe the show's over.' That took quite a while. What was going on?"

"He was crying."

Not too long after that, another friend of mine, Bryce Hand, who had recently left Amherst College for a teaching job at Syracuse University, invited us to show "Suzanne's Lament" at his new school. By that time, we were more or less satisfied with the content of the program, except for the ending. "I really don't know life at all" lacked a little punch, we thought.

Paul and I did the show in Syracuse, and afterward Paul was looking at a Gideon Bible in our motel room when he said, "Hey look at this." He showed me a Bible verse, Genesis 1:28, that made a perfect ending for the show: ". . . and God said unto them, be fruitful and multiply, and replenish the earth and subdue it; and have dominion over the fish of the sea, and over the fowl of the air, and over every living thing that moveth upon the earth." We made a slide with those words on it and used it for the ending of the program.

One of my proudest moments with "Suzanne's Lament" was at a presentation at Norwich University in the small town of Northfield, Vermont, in April 1971. One of our former UMass students, Fred Larsen, was a professor there, and he invited the local populace, in addition to the students, to attend the evening show. Over three hundred people attended, and I noticed in particular an older fellow, maybe in his seventies, who was dressed in overalls and was obviously a local farmer.

As the audience was filing out after the show, the old man shook my hand and said, "Well son, guess it just goes to prove that there's a first time for everything."

Another proud moment occurred after a presentation for the geology department at the University of Texas when my mentor, the great Robert L. Folk, pounded me on the shoulder and said, "I got it M. O. I mean this time I really got it! I got it!!!" His eyes were glistening with excitement.

I just smiled back and tried to escape quickly, thinking "This guy is so creative there's no telling what he's come up with. What would he do if I told him I was only lamenting the passage of a way of life?"

That was the beauty of "Suzanne's Lament"; it could move people in their own directions.

Speaking of changing lifestyles, sometime after New Year's Day 1971, I started cohabiting with a new love partner, Leita Jean, a high-spirited New Englander of Italian ancestry.

In late November 1972, Leita Jean, Frank J. Raffaldi, and I headed for southern New Jersey to show the slide show at one of the universities there. Our host for the occasion had already seen the show numerous times, eventually at least fourteen times, because he was in love with the idea of living in Alaska. But, unfortunately, a wife and other family ties kept him in New Jersey.

Anyway, unbeknownst to me, he had planned a grand reception at his palatial mansion before the show, which was scheduled for 8:00 p.m. We got off to a late start and drove straight to New Jersey from western Massachusetts. Consequently, we arrived at his house famished. As the other guests started to arrive, dressed to the hilt, we sat in the living room scarfing down the cheese and crackers. We were wearing what we usually wore for shows at universities. I had on khaki field pants and a plaid shirt, and Leita Jean and Frank were both dressed in blue jeans.

The stars of the show had just finished cleaning out all the hors d'oeuvres within reach, when we looked up to see the hostess come sweeping down the extensive spiral staircase in a long white evening gown. That was the only time I remember Leita Jean ever saying something about being underdressed for an occasion. With nothing better to do, Frank and I kept gobbling the cheese and crackers that the maid served, while Leita Jean fidgeted nervously.

Eventually, that ordeal ended, and we headed for the auditorium on campus where we would present the program. The auditorium was a little small, but it had two opera-style levels. The place filled up quickly, standing room only. Before the show, our host introduced me to a fellow faculty member, about forty-five, who had a very earnest look on his face.

He said, "I hear you are going to give it to them!"

"Give what to whom?" I thought. I finally figured out that he thought this show was a rallying cry for environmentalists, something that I had never intended "Suzanne's Lament" to be. But, as I said before, people could make whatever they wanted to out of it.

Anyway, I smiled in response to this remark and escaped back to the slide projectors. During the show, I sat behind the projectors, which were set up in the middle of the auditorium. I changed each one of the over three hundred slides in the show in synch with the music. Frank managed the sound system, and Leita Jean handed out the programs. Frank and I didn't let anyone else sit in the row I was in so we could get out of easily to deal with any emergencies with the equipment. After testing the sound system one last time, Frank sat in the same row beside me.

The guy that wanted me to give it to them, let's call him "Agenda," sat right in the front row exactly opposite one of the oversized speakers. Frank had the sound pretty loud, as he usually did for university audiences, especially one that big. In the middle of the first song, old Agenda jumped up and yelled, "The sound is too loud! Turn it down!"

Frank didn't budge.

Then Agenda walked back to our row, sat down beside Frank and said, loudly again, "Turn down the sound!"

Frank just shook his head, as if to say "I've already done this about a hundred times, and I know what I'm doing. No, I will not turn the sound down." Meanwhile, he was staring straight ahead.

I kept changing the slides. Frank was in charge of the sound.

With that, Agenda walked to the back of the auditorium and yelled, "Well then, I refuse to participate!"

Then he left, never to be seen again by me or Frank.

At the end of the show, we had the same warm reception that we usually got from university audiences. I suppose they thought the sound was just about right, as Frank well knew they would.

By the time we did that show in southern New Jersey, we had shown it to over two hundred audiences, with more to come on a big western tour in the spring of 1972. Frank and I visited a number of the major universities in California, the University of Montana, the University of Illinois, and several others. When we returned to UMass, a huge banner was hung across the hall that said "Welcome Home Frank." My office, which was rather large for an associate professor's, was stuffed from wall to wall with waste paper.

In total, well over ten thousand people were in the audiences for the presentations we made in those first two years. Paul and I eventually hired a company to make the slide show into a movie, which we sold to a number of universities. I have no idea how many people saw the movie.

Possibly the most memorable show of all was at the conference of coastal geologists held in Binghamton, New York, in September 1972, at which we presented the paper on the zonal study. I believe everybody in the conference came to the evening show of "Suzanne's Lament," which was preceded with lots of beer and wine drinking. After finishing the show, I walked up front as usual to ask for questions, and when I turned around to look at the audience, I could see that there wasn't a dry eye in the house.

Thinking back on the Binghamton show just a couple of years ago, I wrote in my diary, "That was when everybody cared. Private property wasn't more important then. Jobs weren't more important then. Free international trade wasn't more important then. The earth was in every geologist's heart then. Lying about the facts to make the client happy so it would be easier to buy one-hundred-dollar bottles of wine wasn't a way of life then. The Spirit Father was in charge then. God, I miss those times."

The Seven Ages of Geotechnical Man

As I said before, I don't think I "went crazy" in the summer of 1970, but sometime between the early summer and late fall of that year, I did undergo the "Third Conversion" from the "Age of Arrogance" to the "Age of Paranoia." At the time I didn't know what was happening to me. It took me about ten years or so to figure it out, as the following discussion, which jumps ahead about eighteen years, attempts to explain.

Wednesday, 3 February 1988—
Boulder, Colorado

During a visit to the Boulder office of RPI International, Dan Domeracki, the operating officer of the company, told me that a few weeks earlier he got up at 3:00 A.M. and walked around Boulder in the snow for hours. During that walk he decided that RPI was "his thing." Before that, he had been ready to quit. After all, we had been fighting this losing battle (to keep the company solvent) for almost three years.

He said, "The write-up you sent on the ages of man helped."

He was referring to a chart in which I had tried to show him where he fit into the evolution of the geotechnical man. I had just turned fifty-three when I created the chart, and Danny was in his early thirties.

"I would like to avoid the 'Age of Arrogance,' if possible," he pointed out.

"I doubt if that is possible," I countered and then went into some detail on how I had avoided neither the "Age of Arrogance" nor the "Age of Paranoia," and how living through both of those ages had caused me to take a number of wrong turns that I now wished I had never taken.

I had been thinking about my theory on the ages of geotechnical man for several years, and I had used it often to guide my management policies in supervising a horde of graduate students. The notes that I had given to Danny were the first that I had written about these ideas.

Seventy-two graduate students finished graduate degrees under my supervision at the universities of Massachusetts and South Carolina. Only seven of those students were women, so let's just say that my sample was too small for this theory to apply to women, and that it applies only to men, for the time being anyway. Of course, lots of other colleagues, associates, and acquaintances in the geotechnical world also provided input. I don't know how many other professions that this theory applies to, but I'm pretty confident that it applies to the legal profession, one that I had considerable dealings with as RPI International folded.

According to my hypothesis, geotechnical man passes through seven ages, each separated by either a "conversion" or an "inversion":

1. Age of Innocence: This age ranges from birth to 12–15 years of age.

    Horizons are somewhat limited, geotechnically, and learning is rapid. The age ends at the "First Conversion," when the person is converted to a cause, sometimes driven by the desire to belong to a group. This group could be a religious organization (my situation), the Boy Scouts, the drug culture, and so on, depending upon the environment in which one grows up.

2. Age of Wonderment: Ranges from First Conversion to the late twenties or when the person gets his Ph.D., whichever comes first.

    This is the time of the most intellectual growth, in a geotechnical sense. The individual has tremendous learning capability and focuses strongly on his goals, such as graduation, learning a skill, going to heights in a field to which no one else has ever gone, and so on. This age ends at the Second Conversion, when the person looks around and knows that he is the best. He can conquer the world, which has now become his oyster, as it were.

3. Age of Arrogance: Usually ranges from the Second Conversion until the middle to late thirties.

    This age is arrogance personified, which is defined in Webster's New Collegiate Dictionary as, "A feeling of superiority manifested by an overbearing manner or presumptuous claims." This sorry stage ends when something happens to bring on the Third Conversion. Our hero suddenly realizes that he has overachieved, that it is not much fun anymore, and that life has more to offer. He adopts a different set of values and goals, and they don't usually include "family values."

4. Age of Paranoia: Starts at the Third Conversion and ends in the middle forties, hopefully. Some people never get out of some of these ages.

    The new philosophy brought on by the Third Conversion opens his eyes to reality. He is no longer the man that he used to be, and the world may not be his oyster after all. He is now paranoid about a lot of things, which is defined by Webster's New Collegiate Dictionary as, "A psychosis characterized by systematized delusions of persecution or grandeur, usually without hallucinations." At least he is supposed to escape the hallucinations, but he doesn't entirely. This state is a very bad place to be. I know, I've been there. My fondest wish is that, after reading this book, you do not think that I am still there. Anyway, this all ends with something called the "Enlightenment," when the poor victim gradually pulls himself together, sets realistic goals, and moves on. The Enlightenment is brought about by a miracle.

5. Age of Realignment: Lasts from the Enlightenment until somewhere between the late fifties and the middle sixties.

    With new, achievable goals established and old unrealistic goals discarded, this period should be one of his happiest. The individual should have numerous intellectual pursuits with only learning and enjoyment in mind. One would hope that this age would last a long time. I do, anyway, because I was fifty-three when I wrote the ages of geotechnical man and am now sixty-two as I write this book.

    Stage 6 is the Age of Space Walking and stage 7 is the Age of Senility, but I haven't lived them yet.

You are probably wondering if any of this information helped Danny avoid the Age of Arrogance. Nah, I don't think so. Almost nobody does.

While living through these different ages, I had my share of nicknames. I always liked my name, Miles, because it was so unusual that there was never any confusion about who people were talking to when someone called my name. The only problem was, no one ever called me that. The first nickname that I can remember that I had in school was "Murse," short for "Muscles." Murse stuck with me for a long time until I decided to change it. Mo Modulewski was a great lineman at the University of Maryland at that time (another Polish hero!), and since my initials were M.O., I thought it appropriate that I should be called "Mo," too. A little behind-the-scenes work, and that was the name I kept all through high school.

While playing baseball at Berea College, Kentucky, my name was misspelled as "Miley" in the local paper. My coach, Monarchy Wyatt, started to call me Miley, which he continued to do all four years that I played for him. Speaking of great names, how would you like to play for a coach called "Monarchy"?

Back in 1973, not long after I had moved to South Carolina and while I was still living with Leita Jean, we became friends with another professor at USC-East who was born and raised in Sweden, B.J. Kjerfve, and his wife, Lucy. Lucy's parents visited one weekend and invited us to dinner with the four of them.

As we were introduced, Lucy's father, who, shall we say, was a typical piedmont Georgian, took one look at me with my beard and said, "Oh no, a Dumb Swede and a Billy Whiskers both in the same night!"

Thus, "Billy Whiskers" became my adopted name for the next few years.

PART 3

IN BETWEEN

DANCES

On my way home
I asked myself where I'd been.
I'd been traveling through
a dream with no end nor beginning.
I'd been lost in a fog
with no top and no bottom.
I'd been wasting my time staring
into an empty wasteland.
I'd been hoping for a change
that would free my burden.
I'd been saving my love for
one that stayed behind.
I know she'll be waiting
to hold me close to her.
Wish I could hold her
forever and ever.
And get off this bad trip
for the last time.

Billy Whiskers, 28 March 1971, written
on an Eastern Airlines napkin

The Great Alaska Oil Rush

My stay at UMass ended in August 1972, when fourteen of us in my coastal program loaded the grant equipment, the data, and our clothes into U-Haul trailers and headed south into hot and steamy Columbia, South Carolina. At the University of South Carolina, the graduate faculty was a group of misfits from other departments around the country quickly assembled because the dean of arts and sciences had decided that geology was an area in which the university could become first class. In other words, the geology faculty was something like an expansion NFL football team. It was made up of a select number of rejects and people who couldn't get along with their former teams (that is, their former faculties).

Under the guidance of the graduate director, John C. Ferm, the faculty had already figured out how to run a graduate program. Students were judged primarily on their capability to conduct original research. I agreed wholeheartedly with that approach. Furthermore, they welcomed money hustlers like me, because we could support the students' original research. It was a perfect place for my program at that time, and it flourished. Before long, I had fifteen graduate students, two post-graduate assistants, and several technicians working in the program. Money was no object.

As one of my students, and later a business partner, Erich Gundlach, was wont to say, "Nothing but the best for our boys and girls."

Getting all that money required that I travel endlessly, and a measured chaos reigned. On 13 June 1973, I was washing my hands in the men's room at the Dallas airport when I heard an announcement over the intercom, "Would the party losing a little girl four or five years old, wearing a blue-and-white polka dot dress and a brown cowboy hat, please go to the security booth and claim same!" I laughed when I first heard it, not realizing until later that the message was for me.

Sometime in the fall of 1974, I applied for a grant to continue our studies on the Alaskan coast. By some miracle, NOAA saw fit to include me and my graduate students in their Outer Continental Shelf Environmental Assessment Program (OCSEAP), which was an ambitious program set up to provide baseline data on the relatively unknown nearshore waters and shoreline of any part of Alaska that might contain oil and gas prospects. Our first assignment was the south-central coastal area where we had worked in 1969–1971.

I was in Anchorage to attend a warm-up meeting on the science of the Alaskan coast from both an environmental and oil exploration perspective. I missed most of the meeting because of a delay in Chicago caused by a blizzard, just barely making it in time to give my presentation, which I thought went rather poorly. Apparently one of my "friends" from the U.S. Geological Survey thought so too, because he referred to it as "papier-mâché."

It seemed odd to me that geologists would schedule a meeting in a glass-paneled high-rise hotel in Anchorage, Alaska, the site of one of the most disastrous earthquakes in U.S. history, the Good Friday earthquake of March 1964. As I was sitting in the meeting scribbling notes, a famous professor from a large university in the Midwest was telling a cute story about how his boat captain didn't know the tidal range along the shore of the Bering Sea where he was sampling. I guess that meant they got stranded on a tidal flat or something.

I wrote, "This is a lot worse than sitting in the snow in Chicago!"

He concluded by saying, "I applaud the state of Alaska. No place has a better prospecting program, on either side of the pond." Yes, he said that.

"It's good for industry," he concluded. Wonder if the state of Alaska was listening?

After the meeting, I took a cab to the airport. On the way, we passed Phillips 66 and Chevron and Sitka spruce. The spruce seemed out of place among the typical city fringe of service stations, shopping areas, and massage parlors. So did the sound of Johnny Cash singing "Orange Blossom Special" on the radio, talking about getting some sand in his shoes down in Florida or California.

At the airport, construction crews, mostly old men in their fifties, were traveling to and from the North Slope and the pipeline. I saw a broken-up drag line operator in a wheelchair on his way from Fairbanks to Juneau, laughing and joking with a friend. I saw dozens of Texas oil men, so clearly defined in their white shoes and pin-striped trousers, come to Alaska to get out the oil.

"This is Alaska Airlines Flight 66 to Cordova, Yakutat, Juneau, Ketchikan, and Seattle." These words came in muffled tones over the intercom as I adjusted my seat belt in preparation for takeoff. The little brown napkin wrapped around the glass of Coke had a sketch of a B-727 with an Eskimo's head on the tail along with the words, "Nobody knows Alaska like Alaska knows Alaska."

We were beginning our descent into Cordova, flying at about eight thousand feet over Prince William Sound, when we broke out into clear sky. I was engulfed in blue. As I wrote on a piece of paper, I thought, "Who can describe the blues of this sky and water, and the mountains, the most beautiful I have ever seen?"

Trees came right down to the water. There was no sign of human encroachment anywhere. The mountains were laid out in long lines, covered on their crests by snow and separated by the blue of the water. The lineations showed evidence of the faulting that built the mountains through time as the sea floor pushes against the land, long, linear, healed scars. Then they were lost in the clouds again. A brief, exquisite glimpse of pure blue.[18]

I wrote, "Prince William Sound harbors Valdez, the town at the southern end of the famous trans-Alaska pipeline. Some day the tankers will pour in there to take the North Slope oil down to the 'lower forty-eight.'"

The B-720 rolled to a stop in front of the one-room, log cabin terminal at Cordova. It was April 1975; I hadn't seen the place since August 1971. This time I was just passing through but would be back in about six weeks to begin the new research program for NOAA. The same signs were visible: "Welcome to the Great Land" and "Cordova—Home of the Ice Worm."

I loved Cordova. It was a small fishing village of fifteen hundred inhabitants, huddled between two high snow-covered mountains on the shore of Prince William Sound. It has a fine, sheltered harbor. In late spring, the harbor is busy with the beginning of the gill-netting season. The fishermen string nets across the tidal channels of the Copper River Delta in order to catch salmon as they come into fresh water to spawn.

There are no roads into Cordova. The bridge over the Copper River fell down during the Good Friday earthquake of 1964 and had not been replaced. In Cordova, you could see bumper stickers for and against rebuilding the road, implying the salvation of the town was dependent on the fate of the road. Two jets a day landed at the airport, which is located on the delta thirteen miles from town. A ferry sometimes came over from Valdez. The sun seldom shines in Cordova, but on the rare day that it does, the streets are filled with people (and dogs) in a holiday mood.

I was thinking that no doubt the town would change as the pipeline was completed. However, when I went back the following month, after a four-year absence, I was amazed to see that it was still essentially the same. A bar-motel, the Reluctant Fisherman, was new, and the bowling alley had shut down, but nothing else had changed.

With five passengers and the mail safely on board, we took off across the braided network of streams that build the delta and headed for the next stop, Yakutat. The scenery between Cordova and Yakutat is probably the most spectacular in the whole world. A mountain 18,008 feet high, Mt. St. Elias, overlooks the ocean at the head of Icy Bay. The world's largest piedmont glacier, the Malaspina, which is about the size of the state of Rhode Island, abuts the ocean at the western edge of Yakutat Bay. From twenty thousand feet, its convoluted black, red, and brown moraine lines, mixed with the white of the ice, give the appearance of a giant marble cake spilling out of three openings in the mountains. Very famous in the geological literature. Most of the time, however, the glacier is covered by its rain cloud. Annual precipitation over the Malaspina is over 140 inches, roughly twice that of the Amazon jungle. This day was no exception; it was raining lightly as we came down through the thousand-foot ceiling onto the broad, forested plain that borders the shore at Yakutat.

Yakutat is a fishing village whose population was approximately five hundred. I had not spent any time there except to get fuel and an occasional meal at the Yakutat Lodge, which was one of two lodges in town. I would be back soon.

At 5:00 P.M., we went down the airstrip through the woods with a loud roar and then up into the clouds, and all the passengers who got on the plane in Yakutat began the wondrous transformation—from living in the rain, from sampling on the boat, from the enclosed interior, rained-in aura of huts in the rain forest, from the poolrooms in the bars, from the euphoria of alcoholic dreams to the tight-skirted, disinfected man-made capsule of time magic, to the multi-layered concrete electric lighted noisy air terminal in Seattle, to the skyscrapers by the lake, to the Mexican food restaurant with its artificial flowers. And to the honey-scented ladies in the restaurants and to Copper Calhoun dancing in the dark.

"Want a date?"

Seattle, City of Concrete

First Meeting

All the way down there
On the tracks
We walked
And talked some.
We held fast and looked
At the trees,
And perhaps the stars,
I don't remember.

Billy Whiskers

Saturday, 5 April 1975—
Seattle, Washington

As we approached the airport, I looked over a surface of ruffled clouds that covered the lowlands surrounding the city to where Mt. Ranier, a huge volcanic mountain, jutted up into the blue overhead like a Mayan temple or Egyptian pyramid with a foamy mat of white clouds around its base.

At the airport, I caught the shuttlebus for downtown Seattle, reading the signs along the way, "Tacoma Exit," "Tukwila Exit," and "Boeing field, second right." Concrete and painted lines; bridges over the road. "Stand still and watch the world go by and it will," from a sign in a churchyard. "Jesus is coming soon," a little homemade sign on top of a small house set back from the road. I'm still waiting. Ha, ha.

About 3:00 P.M. the next day, I was walking down Pike Street near Third, and I saw a young man playing a guitar and singing. A small group of teenagers had gathered to listen. An ancient derelict was prancing around him blowing away on a harmonica.

The guitar player was skinny with a slightly scared face and curly hair. He wore jeans and tennis shoes and as he whanged mercilessly on the strings, he wobbled to and fro a little bit, lifting his right tennis shoe up and down. A small cap was turned upside down on the concrete, a dollar bill, a few coins, and a note inside. He was singing a song about tourist towns, tourists, and their trash, and about how they should all have to live out their lives in their Winnebagoes. The kids giggled at me as I approached because if ever there was a tourist, it was me, in my sight-seeing costume and with a two-hundred-millimeter lens around my neck.

I had walked by but stopped and returned when he started his next song, about people going up to Alaska to mine the black pipeline gold in what he called the "Great Alaska Oil Rush." I spoke with him briefly and dropped a bill and a note in the cap.

He later wrote to me, "Here are the lyrics you asked for. I hope they will be of use to you. This is a very popular song here—popular and controversial—since this city is an outpost on the way to the pipeline." His name was Jim Page, people's ballad singer.

I lay awake for a long time back in my hotel room, thinking about the rest of the trip. I was headed for Dallas, by way of San Francisco, to attend the annual meeting of the American Association of Petroleum Geologists, where I would give another speech. This speech described how the sand islands along the east coast, and elsewhere, were created. I presented my drumstick model, explaining how barrier islands are shaped like chicken legs, profusely illustrating my points with pretty pictures of beaches, islands, waves, crispy brown drumsticks, and Colonel Sanders. From Dallas, I traveled to Memphis and the annual meeting of the southeastern section of the Geological Society of America.

In Dallas, I met Leita Jean, who was no longer my live-in friend but had helped me with the paper, and spent a few uneventful days playing Dr. Science. Then I went on to Memphis alone, where I visited the Schlitz brewery, got drunk, and lived in a tall Holiday Inn that sits on the bank of the Mississippi River and is shaped like a Schlitz beer can. I also listened to papers by aspiring graduate students and young college professors and asked obnoxious questions if I didn't entirely agree with the speakers' points of view—the first and only time in my career that I ever did that. Maybe it was because the presentations were so sparsely attended, usually by less than twenty people. Therefore, it was kind of like a thesis defense. And nobody was passing!

I also had fun spending the last of my traveling money on dancing girls.

On the morning of my last day in Memphis, I was standing by the Holiday Inn window looking out over the Mississippi River. The river was high and on the rise. I had seen farm houses under water as the plane landed a couple of nights before. It was early spring in Memphis, and the sun was bright on a clear day. A long freight train was crossing the trestle over the river, and a river boat flying an American flag was just passing under it. The green tops of the budding trees were a strange sight as they protruded up through the muddy water of the river.

Later, as I was sitting in the plane at the ramp, I remembered that this was the last leg of the trip. I felt a tinge of regret, perhaps some resentment, that the trip was over and that I would have to go home. To what? To the empty house? To the dog? To Peggy, who had already left the office? To the unpaid bills? To the messages on my desk? To the unread dissertations? To being boss?

Then once again, I felt the slight tug along my spine as I was forced back into the seat as we started down the runway, and then the peculiar light feeling in the top of my skull as we lifted off the ground.

"Ladies and gentlemen. We are now on our approach into Atlanta. Please remain in your seats with your seat belts securely fastened about you."

"Welcome to Columbia, home of the fighting Gamecocks!"

The rest of the month of April was passed in Columbia in a flurry of activity, getting ready for the summer's research in Alaska, reading theses, and wrapping up departmental business for the academic year. I spent a lot of time working on a new slide show on Iceland called "A Day Just Like Today,"[19] which I showed to a gathering of friends on the first Saturday night in May.

In the show, I used the same technique we used in "Suzanne's Lament" to describe the unique experience of working in a remote and beautiful place, namely by letting the songs I had selected carry the story. I showed pictures of Leita Jean and my new girlfriend Debbie, both of whom were in the audience, and my two young daughters mixed with slides of Impressionist paintings in a sequence set to the words of "Just Like a Woman" by Bob Dylan. That was my favorite part of the show, but some people felt it was out of place in a nature story. Afterwards, we had a celebration at my house with lots of dancing and carrying on.

On the Sunday afternoon following the party, Jon Boothroyd and I headed back for Seattle to finalize the paperwork on our Alaska grant. Upon completion of that tedious activity, we returned to Columbia via Omaha, St. Louis, and Atlanta.

As we approached the St. Louis airport, I thought: "Somewhere once in my distant past, I lived in St. Louis for two years. I can see the lights of the city as we glide in. That was a long time ago. The second year Barbara Anne and I lived in that small apartment. And we sometimes went down to visit the beautiful rivers in the Ozarks. They gave me an M.S. degree for a piece of crap. But I liked the city, the late movies, the softball games, and the beautiful girls in the labs. And the old fashioned streets, the street cars, and the park. And the Cardinals. I saw Stan "the Man" Musial, my childhood idol, hit a few high into the stands, although he was as old then as I am now.

"'We don't need a flood; we need a tidal wave,' an old man yelled as the rookie Curt Flood came in to pinch-hit for the Cardinals. We had been bombarded with Pittsburgh home runs all afternoon in the left field bleachers.

"I loved Maria Schell and lusted after Bridget Bardot in the late movies. I got drunk for the first time, and I learned how to teach under a master. I rode on the bus with Barbara Anne back to Shoals, Indiana, which was halfway.

"'If you're looking for a flophouse you won't find it in this town,' the old man at the desk told me as I was looking for a cheap place to stay in my thesis area in western Missouri. The St. Genevieve limestone and the St. Peter sandstone and a C in optical mineralogy."

Back in South Carolina, I drove to the coast on Thursday morning to meet my field crew and have one last look at Kiawah Island before going north. Our project to draw construction set-back lines for a Kuwaiti-owned development company was drawing to a close. The application of those set-backs would become something of a model for such developments in the future.

Leita Jean was working on that field crew, putting in the last day of field work she would do for me that summer. It would be strange to be going to Alaska without her. For the past four and a half years she had accompanied me in the field on many of our major projects, from the snow-white sand beaches around Panama City, Florida, to the red sandstone cliffs of Prince Edward Island, Canada. But the next day she would go to Florida with her new lover, to run on the beach with him during the day and serve cocktails to Holiday Inn travelers during the night.

Holiday Inn

It's built
like a big
beer can, and,
as always,

sits beside
a superhighway.

On top
is a round
restaurant and
a round bar
with a tall
brown girl
serving drinks,
a fat
round man
singing songs,

And a
middle-aged waitress
handing out
TV dinners
to the
tired travelers
seeking refuge
from the
heat of the road.

I see an
exit sign
mirrored against
the sky
as I look
away from
the fountain
of voices
mixed with
the guitar sounds.

And I see
the moon
almost full
fuzzy behind
the light smog
and glistening
on the river.

The whisky sour
is too sweet
to drink
but I sit
and try
to think of
a clever way
to get
her attention.

Because I really
don't want
to go back
to the TV
and newspaper
in a pie-shaped room
with fake bricks
fake furniture
dirty windows
and a
sanitized toilet bowl.

The highway below
is full of cars
heading for Tampa
buzzing by
the no vacancy
sign continuing
in search of
the next
Holiday Inn.

Billy Whiskers

Back at the university, we made last-minute preparations to begin the
field work. We would split up into two field crews once we got to Yakutat. Jon Boothroyd and his crew of two, Ray Levey and Mark Cable, were
stationed in a camp by the Malaspina Glacier where they focused on the
dynamic processes of streams that drain the glacier. My team, which
covered the entire shoreline between Cordova and Yakutat, consisted of
myself, Chris Ruby, a graduate student working on a M.S. thesis on the
project, and Janie Zenger, an enthusiastic undergraduate English major.

Cordova

The summer field season started in Cordova, where we carried out our work by landing a small aircraft, a Cessna 180, on the beaches. Once on the beach, we conducted topographic surveys and collected sediment samples. We were covering approximately three hundred miles of shoreline. One of the most striking aspects of the area is the variety in the landscape. In places, the mountains come right down to the beach, and in others, a wide plain or a massive glacier separates the mountains from the sea. In general, the weather is cool and damp, but the exceptional clear day makes up for the rest because the sight of eighteen-thousand-foot, snow-covered mountains as a backdrop for our field surveys was an inspiration. Unlike the shoreline of South Carolina, where the vegetation is pruned by salt spray, huge fir and spruce trees grow right to the water's edge. In places where the shore is eroding, they are toppled in the surf at high tide in a chaotic array of intertwined roots and trunks, giving extra beauty to the shore even with their untimely deaths.

Out of Cordova, we worked on the Copper River Delta, which is a complex array of islands, channels, and tidal flats. I wrote the following in a letter to my friend Sally:

The truth came to me in this vision I had Friday morning, a spectacular clear day, as we were flying at five thousand feet taking pictures. The truth is that barrier islands really are shaped like drumsticks, and as the drumstick moves, the knee joint grows at the expense of the thigh joint. Anyway, the pattern of island change and development is super clear at five thousand feet on this one island, Egg Island, probably named that because it is inhabited by thousands of sea gulls.

There have been some remarkable changes of the shoreline since 1971. So far, we have been lucky enough to find several of our old surveying stations, so we are getting good data on the changes that have occurred since we were last here. At one point on Egg Island, the shoreline has built out into the ocean over one-half mile in the past four years. Along the front of the Bering Glacier, intensive erosion has occurred, as is evidenced by the presence of huge erosional scarps.

I went on to tell her about our work schedule, our day usually starting about 3:30 A.M. to catch the early morning low tide, with a break in the middle of the day and another trip out in the afternoon to work the afternoon low tide. I continued:

From about 9:00 P.M. until we quit, we circulate among the Club Bar, the Alaska Hotel Bar, and a place we call the Fisherman's Dement (the Reluctant Fisherman), to drink this fantastic beer they have up here called Olympia ("Oly" for short) and admire various and sundry bottoms and belly buttons and play pool. Anyway, this is the schedule we have been keeping for the past few days. It will change as the tide changes and we move from place to place.

Actually, the dominant theme here is the spectacular beauty of the place and the joy of being away from the phone and other hassles that go with being back there. The science is good also, and I have been getting a lot of ideas for papers to publish and new research projects.

I miss you, even though you are terribly grumpy sometimes. But, your loving is fine, and I like your stories.

Has the mad priest created havoc yet? Has the university fallen to ruin in my absence? Have you seen my dog? When I get to Yakataga, away from the bright lights of Cordova, I will try to write something serious about nature and all that.

Monday, Memorial Day, 26 May 1975 —
Cordova, Alaska

I had been sleeping fitfully, dreaming about flunking out of college, various weird pets, and other strange things, when I sat up to look at the clock again. It was 4:20 A.M. The clock was set to go off at 4:30. The sun had been shining through the dim overcast for some time. I got up and dressed; while I was in the bathroom, the clock woke John the Pilot, a tall, handsome mystery man who said very little and slept at every opportunity. By 4:45, we were on our way out the dusty road toward the airport. John arrived before us, and by the time we got there, he had taxied the plane to the end of the narrow gravel strip that sits precariously on the side of the mountain by Eyak Lake. The wind was blowing strongly as we took off, winding our way between the mountain peaks and out over the Copper River Delta.

As John called in our flight plan to the Cordova Airport, the weather report came back over the radio, all bad. We decided to go ahead anyway. We were heading for Strawberry Reef, a barrier island on the eastern end of the delta, to resurvey a beach-profile station that I had established in February 1970. A beach profile is a measurement of the intertidal surface (perpendicular to the beach) in order to both describe the beach shape and to determine how it changes over time.

The wind was blowing out of the east-northeast. We later measured it at gusts of 40–45 mph.

When we reached the shoreline and started flying toward Strawberry Reef, I could see sheets of sand driven along the beach by the wind. The entire surfaces of the islands were covered by peculiar flattened-out dunes with wavy linear shapes oriented perpendicular to the wind. Long fingers of sand streaked between the dunes and over the intertidal flats.

We landed on the beach by the survey stake we had located earlier from the air. The orange stake was barely visible in the blowing sand. The wind was cold, and the sand bit into the backs of my legs as I tried to put on my waders. Drops of rain were mixed with the flying sand.

Out over the ocean, a low, black overcast came down against the faint white and green of the horizon. The surface of the ocean foamed white as the waves chaotically bristled in the wind. Looking east down the beach, I could occasionally see the dim outline of Cape Katalla, a rocky headland at the margin of the delta. Isolated clouds seemed to sweep from behind the Cape and dart over the delta. The sky was a kaleidoscope of grays and blacks. Gray-white light rays from the sun sometimes penetrated the clouds and fanned over the delta, like searchlights scanning the surface of the tidal flats.

As we struggled up the beach into the wind toward the profile, I turned to walk backward against the wind. Looking back down the beach, I saw the plane centered in a half halo of a perfectly shaped rainbow, one end dipping into the breaking waves offshore and one end into the dunes on the island. I ran back to the plane to get my camera because it was the first time I had ever seen such a perfectly formed rainbow on the beach. Every time I prepared a slide show set to music, I found songs I wanted to use that contained the word *rainbow*, but I didn't have a single picture of a rainbow. So I took about ten pictures, and then we went on to measure the profile.

John, who waited behind in the plane, said the blowing sand obscured our legs below our knees as we walked across the dunes, giving the appearance of other-world creatures moving on a planet made of some gas-like substance.

The profile was about one-half mile long, and on each measurement, we fought with the tape against the wind. I was sighting on the horizon to level between positions on the profile, and I had trouble distinguishing between the green ocean surface and the black sky. We moved slowly. The rainbow brightened and darkened, depending upon how much sunlight broke through the clouds. My hands turned numb with cold, and my fingers locked around the profile rod.

Standing in the surf at the end of the run, I felt caught up in an orgy of energy and light. Looking toward the sun, I could see the light bouncing off the water that was being blown randomly in sheets across the bar surfaces. Waves bending across the bars bumped clumsily into each other as their curling breakers collapsed against the fury of the wind. My body was chilled to the bone, and the wind seemed to blow right through it, as if it were a fixture of the beach itself.

On the way home, we passed over some seals on the bank of the river and saw geese and ducks huddled low in the channels. We heard on the radio that one of the fishing boats had swamped in the surf and that one man was washed overboard. A survivor was plucked from the boat by a helicopter. One dead, one alive. We landed on the strip where the ambulance waited. As we walked up the road, the helicopter came in, and the ambulance drove by with its lights flashing.

At 10 A.M., we were eating pancakes in the Club Cafe.

Sad Song of the Fisherman

Where white clouds pile high on jagged peaks
And the green pond stirs with ripples
In this place I hear a fisherman
Now and then dip his oar and sing:
His voice, and then his voice again,
until I cannot listen;

It makes my thoughts too sad . . .
That is, though the fisherman bears
me no ill will, his songs poke
holes of sadness in me.

Han-Shan, *Cold Mountain*[20]

Thursday, 29 May 1975 —
Yakutat, Alaska

We were working out of the Yakutat Lodge for a few days as we waited for the remainder of our gear to be shipped from South Carolina. It rained almost the entire time we were there.

The next morning, we flew around the front of the giant glacier with black cliffs hiding the ice underneath the trees. Sea lions sat on the big boulders at Sitkagi Bluffs. We luckily found profile Mal-3, marked by a bent stake on a small dunehill in front of a river, and surveyed it. Everything looked whiter and grayer than I remembered. Bear tracks were everywhere. No plants were out yet, except for the young strawberries that were all over. Then we went across Icy Bay to station YKG-3. I still got goosebumps when we flew across those big bays in a single-engine

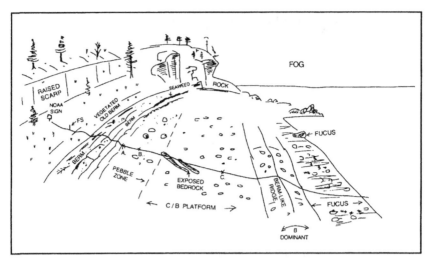

FIGURE 2: Sketch of a beach station in Alaska.

airplane. Profile YKG-3 brought back memories of when we landed there in February 1970, on a beautiful clear day with snow on the mountains, and took the pictures that we had used in "Suzanne's Lament." It looked drab in comparison on that gray, overcast day in May 1975.

Early the following evening, our tenth day in the field, Chris and I were walking along the road between the Glacier Bear Bar and the Yakutat Lodge, confident that we would soon get a ride. See, the Yakutat Lodge was five miles from town, and the Glacier Bear was halfway in between. It was easy to get from one place to the other because all you had to do was walk along the road and someone always picked you up. Someone was always driving on the road because it was the only one in Yakutat.

It wasn't hard to tell from the smell of the mutilated green Dodge van that its driver must be a fisherman. He was about thirty-five and was big, six feet two inches, with a large belly, long hair, and a long, black beard.

"Where are you from?" I asked.

"Alaska."

"No, but what part of Alaska?"

"Oh, Alaska."

Then he looked out toward the mountains as we drove quietly down the road for a few minutes. The sun was setting, and on that rare clear day, we could watch Mt. St. Elias change colors in the distance. The shadows cast by the giant trees beside the road and their stillness in the fading light created a sublime and peaceful feeling as we drove along.

He started talking again. "Yeah, I've lived in all these towns, the little ones, the big ones. I like being in a middle-sized town like Yakutat the best, though.

"You guys are from South Carolina, right, and I've been wanting to talk with you. I don't like your landing on my property."

This startled us a little bit, but by then we had arrived at the Yakutat Lodge, so we started drinking with him. At first he told us he was a miner and that we had been landing our plane on his claim on the beach south of Lituya Bay. He wanted us to tell him how much the beach was eroding. Then he brought in a big plastic bag of black sand from his van and put it on the table, suggesting that we do a grain size count on the sand for him, and while we were at it, see how much gold was in there. He was great at double-talk, using big words or coining his own. During the winter he was a plumber and worked on the North Slope, making a fortune, or so he said, and I didn't doubt it. In the summer, he fished for halibut by setting trot lines all over Yakutat Bay and in the Gulf of Alaska. The locals called him "Halibut John." He had his own boat. He dabbled in mining on the side.

He eventually decided that there would be more action down at the Glacier Bear, so we loaded into his van and headed for the Bear. John took along a little cherub who had been sitting at the bar whom we later came to know affectionately as "Mr. X," the man building the Road to Nowhere.

The Glacier Bear sat on the corner of a logged out area. Gross-looking piles of stumps, dirt, and logs stood out above the Bear, as if they were keeping watch. The building had no windows. It had a bar area, a dance platform, and a number of tables. When we arrived, the place was teeming with people, mostly very drunk, dancing wildly to the sound of a two-man band imported all the way from Anchorage. "Shorty the Baggage Man," who worked for Alaska Airlines, was so drunk he could hardly stand up, and his attractive wife was dancing with everybody. Another young woman, married to an old white man who sat placidly looking on, also danced with every man there, except Chris, Billy Whiskers, and Halibut John. Finally, Chris and I took the plunge and asked two teenage Indian girls at the table near the platform to dance, and they turned us down cold. Halibut John, who was sitting at another table with an ancient Indian lady, said, "If you struck out there, boys, you might as well go home."

Which we did, not long thereafter. However, we didn't leave before John asked the girls, and they turned him down, too. An old Indian man who had grabbed Janie in the Glass Door, a bar we were in the afternoon before, was dancing all over the place all by himself, also drunk out of his mind.

When we left, it was almost dark. The road was cool and quiet. I could see Jupiter through the treetops and the dark shadow of Mt. St. Elias on the horizon. We had just reached the main road and turned toward the lodge when a pickup truck pulled up and started to turn in the opposite direction. It was S. B. Mann and his wife. They offered us a ride to the lodge. She was driving. We climbed up into the bed of the pickup, an ensemble of four drunks if I ever saw it. Shorty had almost passed out. As Mrs. Mann drove wildly down the road, I leaned over the open window of the left front door and carried on a wise conversation with her. We were contemplating dumping Shorty in the swamp.

The next evening we were eating dinner at the Lodge, and Halibut John joined us. Afterward, we sat in the bar and talked for a while. He told us a lot more about his life. He showed me a picture of his two children in Anchorage; then he quoted verbatim a poem by Robert Service about the mining and the beauty of the Alaskan land. The words were:

I wanted the gold, and I sought it;
I scrabbled and mucked like a slave.
Was it famine or scurvy—I fought it;
I hurled my youth into a grave.
I wanted the gold, and I got it—
Came out with a fortune last fall—
Yet somehow life's not what I thought it,
And somehow the gold isn't all.
No! There is the land (Have you seen it?)
It's the cussedest land that I know
From the big, dizzy mountains that screen it
To the deep, deathlike valleys below.
Some say God was tired when He made it;
Some say it's a fine land to shun;
Maybe, but there's some as would trade it
For no land on earth—and I'm one.

Robert Service, "Spell of the Yukon"

Friends and Lovers

Monday, 2 June 1975—
Yakutat, Alaska

The sun finally came out, and we went south of Yakutat to sample along the shore of the Yakutat Foreland. We did five sampling stations, working a midday low tide. The fine weather put us all in good spirits, exhilarated to be out there with the mountains and the clouds. We stopped at the Dry Bay Fish Camp and visited with the lady in the kitchen, wife of one of the fishermen, from a small town near Bellingham, Washington. I said, "Mary, you've got a great view from your kitchen door!"

You could see over the braided channels of the Alsek River to the snow-topped mountains behind it. The pantry was open, and they had an iron stove. They would start fishing the next day. She gave us free cake and coffee. I felt a little kinship with her and John the Pilot that day. It was a beautiful time.

As we stayed on in Yakutat, our list of friends grew, especially Janie's, who had a knack for meeting people. We even came to know Mr. X better. He was an engineer for the state of Alaska sent to Yakutat a year ahead of the construction crews to build the Road to Nowhere—all by himself, I presume. The Road to Nowhere was supposed to go into the woods on the Yakutat Foreland. The goal was to eventually cross the mountains and link up with the ALCAN highway. Mr. X sat at his desk in the Gulf Air Taxi for hours looking at plans for the Road to Nowhere. He didn't expect to ever reach the ALCAN. He didn't.

We were discussing him one evening at dinner and concluded that Mr. X was the perfect man to be building the Road to Nowhere. He was short, built close to the ground, about forty-five, with an idyllic smile and innocent look, drinking coffee at the cafe and Oly at the bar. Chris said, "He's up here for some other reason."

When we walked outside, he had a big Steelhead trout in the back of his truck, bringing it to George the innkeeper to have it cooked for his dinner.

Next morning, I was walking back toward the dune ridge taking pictures, and I could see the fir trees on the plain stand out in sharp contrast to the soft angles of the high, white mountains in the background. I said to Chris, "Anybody that lives in the East has to be crazy!"

And I decided to move from South Carolina and come to Alaska to live out the final verses of the sad song of the fisherman. That was one of about fifty times that I have decided to move within the past twenty-five years.

"Damn, those mountains are as beautiful as hell!" said the new arrival in Room 6.

That night I received a card from my youngest daughter, Mya Suzanne, that said, "Dear Dad: We are going to be there soon. Dad, school is out June 19, 1975. Dad, we are riding down to N.C."

"Does 'Both Sides Now' have a special meaning for you?" Janie asked. We were alone in the hut a little later that same night. Chris was over at the bar drinking Oly with Mr. X and Halibut John. I thought she was reading.

"Why?"

"Because of the way you sang it. It was beautiful."

"Yes, it does."

Jeff Brown had sent us his guitar in the gear from South Carolina.

Over the next few days activity showed a marked increase. John the Pilot quit and was replaced by another pilot, Gayle Rainey, who needed a few days to get used to landing on cobble beaches and low-tide terraces, with a landing in the ocean being part of the training procedure. We moved on to Cape Yakataga, where we lived in an abandoned FAA station. We were sad to leave Yakutat and all the friends we had made but looked forward to the isolation of the Cape.

The last mail call before leaving Yakutat had a letter from my new friend Faye, a stewardess for Northwest Orient Airlines. We were planning another rendezvous in Anchorage soon. In my answer to her letter before we left for the Cape, I quoted from a poem in *Cold Mountain*, changing some of the words to fit the Alaska setting.

I look at far off St. Elias' summit,
Alone and high above the crowding peaks.
Spruce and fir trees sing in the wind that sways them;
Sea tides wash beneath the shining moon.
I gaze at the mountain's green borders below
And discuss philosophy with the white clouds.
In the wilderness, mountains and seas are all right,
But I wish I had a companion in my search for the Way.

Then I wrote:

But how long will this beauty last? There's a blight creeping up the West Coast of North America from Oregon, Washington, over Vancouver Island, and now it's moving into this unique and beautiful place. It's called "clear-cutting." When a logging company clear-cuts, it takes out everything. All that is left is the bare ground and piles and stumps that sometimes may or may not get burned. Flying over Vancouver Island you can see entire mountainsides completely skinned of timber that have since developed erosional topography that rivals the Badlands of South Dakota. But this procedure is sanctioned by the U.S. Forest Service, and in the Forest Service–owned lands around Yakutat, clear-cutting is in full swing. These cleared zones around Yakutat will not likely have any long-term deleterious effects because the landscape is flat. However, at a new logging camp at Icy Bay, trees are being clear-cut on slopes that I would estimate to be thirty degrees. All this cutting has taken place in the interim between my last visit in 1971 and now!

Yesterday I saw the trees by the river's edge
Wrecked and broken beyond belief,
Only two or three trunks left standing,
Scarred by the blades of a thousand axes.
Frost strips the yellowing leaves,
River waves pluck at withered roots
This is the way the living must fare.
Why curse at Heaven and Earth?

Han-Shan, *Cold Mountain*

I was cursing at heaven and earth when I saw the new logging at Icy Bay, because I knew the trees were going to Japan. I also knew they were being sold at cut-rate prices and that the byproducts would be shipped back to the United States at a huge profit. I also knew the land was not private but was owned by the state of Alaska.

"We don't want you ____ [expletive] from the Lesser Forty-eight coming up here telling us how to use our land!" an irate drunken fisherman had yelled at Chris in the bar of the Yakutat Lodge the night before.

So it goes.

Field Days—Cape Yakataga

When people see the man of Cold Mountain
They all say, "There's a crackpot!
Hardly a face to make one look twice,
His body wrapped in nothing but rags . . ."
The things I say they don't understand;
The things they say I would not utter.
A word to those of you passing by—
Try coming to Cold Mountain some time!

Han-Shan, *Cold Mountain* [20]

Thursday, 12 June 1975 —
Cape Yakataga, Alaska

We had been at Cape Yakataga for three days. The work was a straightforward routine, but we still had a lot to do before finishing the basic sampling. We were isolated there, rarely seeing anyone but our own field party. The mountains are exceptionally imposing at Cape Yakataga; they are only a mile or two from the beach.

This day started out as an especially nice one: no rain, sixty degrees, and only lightly overcast. I was trying to observe, to absorb my surroundings, so I could remember what it was like on the beach and write about it in my diary when I got back to camp.

I wrote these words on the back of a data sheet: "Back in my room, I can't hear the constant roaring of the waves, feel the coolness from the snow up on the mountains, breathe the musky odor of the seaweed thus exposed on the rocks at low tide, nor get the mood back again that the sea has dropped over me like a quilt of somber reflection."

I thought about the ancient glacial till and varves, which are glacial lake deposits, exposed in the rocks under my feet. Such a beautiful pattern! Much like the rocks we had looked for so hard in the Permian strata in Antarctica a decade earlier. This glacial deposit, just like the ones we were flying over every day, was fifty million years old, pushed up and plastered against the older mountains by the magical conveyor belt that runs under the ocean.

Then I walked back to the beach and started sketching that hopeless jumble of rocks, sand, boulders, and mangled trees somebody called "Umbrella Reef," which is located halfway between Cape Yakataga and Icy Point at the western margin of Icy Bay.

Toward the end of the day, it rained, and the wind was cold. As I huddled under the double cover of my Antarctica jacket and rain slicker, the thought occurred to me that both of my ex-true-loves, Leita Jean and Debbie, were basking in the Florida sun with their current true loves.

Three days later, 15 June, we worked five stations, on both sides of Icy Bay. Gray streamers of mist hung low over the mountains. Down on the beach, I felt cold on my hands and spray on my face, wetting my lips. Patches of blue and gray reflected off the water. The beach was a cold, flat, untouched surface, marred only by my own footprints. Then I put my foot on the flattened pebbles of the upper beach, trying to push the impression through my brain that I was there, then. And that next month it would be like a dream as I walked on a beach in South Carolina or some other place yet unknown. I wanted to keep it always fresh in my mind, as the gulls squawked above me.

On the following Saturday, we launched another early morning start. I was singularly impressed by the accumulation of dead trees on the washover terraces east of Icy Cape, at the mouth of Icy Bay. The trees were scattered across the terraces like a bone yard of dead dinosaurs in ordered rows, their roots pointing into the wind, toward the southeast, as if warding off the harshest blows from the waves. Apropos of the junk piles in the towns, Yakataga deserted and the dead railroad at Yakutat, in that land of busted dreams.

After we finished our surveys for the day, we went to Yakutat for fuel. We went by Jon's camp at Yahtse River, and he, Janie, and I went on to Yakutat with Gayle. In Yakutat, we picked up the mail and ate some of Halibut John's catch at the Yakutat Lodge. Janie joined some people in the bar. After a brief visit with our friends at Gulf Air Taxi, I went to look for Janie, in my usual impatient way. I was ready to go on back to Cape Yakataga.

I walked into the room filled with people. I couldn't see Janie anywhere. I did see Mr. X with his small sidekick recently imported from Juneau.

"How's the beach?" Mr. X asked.

"Wonderful," I replied.

"Peaceful?"

"Yeah, very peaceful down there."

I was standing in the middle of the room, a little awkward, in my dirty shirt with my long hair flying. I hadn't had a real bath in over two weeks. I asked Mr. X if he had seen Janie.

"She's right there." He pointed behind me.

She sat at the end of a table full of clean cut, oil company geologists. When I walked over, she introduced me to her friend. "Dr. Hayes, this is Tommy Best."

An awkward silence over handshakes.

"What are you doing here?" I finally asked.

In his embarrassed, half-stooped pose, the young man forgot the name of the company he was working for. "Amoco!" from a chorus of voices. Several self-contained smiles down the table. I tugged at my hair, feeling as awkward as the twenty-year-old. Distinguished scientist-bum meets rugged, all-American oil company jocks in the middle of the Yakutat Lodge dining room. Right there in front of the Alaska Air Lines jet and all that.

Janie exclaimed, "They're coming to Cape Yakataga next Wednesday!"

"Come and see us," I said. All around, self-contained smiles again. I touched Janie on the arm, urging her to come along soon, and left.

Outside, I walked across the white concrete runway in the rare sun of a coastal Alaska late afternoon. As I passed the waiting B-720, I looked up at the open door where a black-haired stewardess leaned, musing over the Yakutat scene. I walked on thinking, "Tonight she'll be in Seattle in the arms of her lover, while Billy Whiskers sleeps alone in Cape Yakataga."

Shortly thereafter, we flew out across the bay, the sun playing on the mountains. As we crossed safely to the west side, I looked down on one of the many gravel spits that occur along that part of the shore and thought:

Malaspina Spit

The waves broke
on the spit
as though they
knew
their appointed job
was to move
it down
the bay always

adding and adding
with each wave
a new growth
to the land
the great land
called Alaska.

But on the
other side
of the bay
a different type
of wave
this time made up
of men
from Amoco
and Atlantic Richfield
sat at the table
eating halibut
and salmon
thinking about
their wives back
in Houston
and in Tulsa.

And
a little bit
about how the oil
was made
and filtered
into the sands
and gravel that were
brought to this same
resting place
by the ancestors
of these waves
fifty million years ago.
Braided streams
and gravel spits
now as then
always trying to
conquer the sea.

———————————

Billy Whiskers

Father and Son

Sunday, 22 June 1975—
Cape Yakataga, Alaska

We got up at 3:45 A.M. and were sampling profiles in front of the Malaspina Glacier by 5:00 A.M. It was a dreary, overcast day, and it rained on us a couple of times; however, we were happy with the science, and the work went smoothly. Wildlife was particularly abundant on this day; at every station we saw bear tracks, and at one place, Gayle saw two moose. Sea ducks were flying up and down the beach all day, and seals would pop their heads up out of the waves as we sighted along the profiles. We had only one more week left.

At 10:30, we went over to Yakutat to get fuel and breakfast. We saw a lot of new faces at the Lodge. Ron the bartender had been fired and was now working on the boat for Halibut John.

I wanted to send Faye a letter, so I went over to the air terminal to see if I could find anyone who would be willing to hand-carry it to Anchorage. While I was waiting, I noticed a tall Indian man, about twenty years old, hugging his young girlfriend, who was part Indian and part white. I asked him if he was going to Anchorage.

He said, "No, Fairbanks," as they continued to hug, with several little guys tugging at their legs.

An older Indian man, who looked about seventy, wearing a suit and hat like a doorman's, was pointing to the topographic map on the wall, talking to a young boy about nine or ten.

"Jimmy," he said, as he moved his finger along the map, "we used to hand-row our boats from here," pointing roughly to the location of the mouth of the Situk River on the outer beach, "all the way around to here and here and here."

He moved his finger into the bay and up to the head, to the Hubbard Glacier, and then all the way around to the end of Russell Fjord.

"Uh," the kid responded, showing a slightly embarrassed interest.

The old man was obviously still high from the night before. You could smell the wine from five or six feet away. His top teeth were gone.

The lower gums had a few irregular snags. His hair was pure white. His eyes were partly hidden behind darkened glasses. His words were broken and hard to understand.

"We didn't have outboard engines in those days. Hey, you know how far that is? How much is two times seventy-four?" A pause. "One hundred and forty eight, right?"

"Yeah," the kid reluctantly agreed, looking for an escape.

The old man continued yammering. I couldn't catch all the words, so I moved in closer to hear the conversation better.

Finally, I asked, "Why did you go up there?"

He muttered several incoherent words. I think he said he went in to catch seals and to hunt. Apparently, his father was a hunter.

"Did you ever hunt bears?" I asked.

He said his oldest brother was a hunting guide. "I don't want to brag." I said, "That's all right."

He worked for his brother for three years as an assistant guide, but then, for some reason, he quit. They hunted glacier bears. He looked around the terminal for the mounted specimen, then noted that it had been moved to the Glacier Bear Bar. He said he and his brother were the only ones that knew where to find the glacier bears.[21]

"Did you ever hunt sheep?"

"Yes," he responded and then pointed to several places on the map where you could find sheep.

"Are you from Yakutat?" I asked.

"Yes, I've been here all my life except for one time, during World War II." I couldn't understand all the words. As best I could determine, he was in some branch of the service during World War II.

"I was lucky to get out of that alive." He seemed proud of it. He was a sharpshooter. He talked about the different kinds of guns he used in the war. Then he said he was discharged out of the Quartermaster Corps.

"Did you fight the Japanese?" I asked.

"That's one thing I don't want to talk about," he said, shaking his head.

"Okay. Don't want to think about it, eh?"

"No. They're human just like you and me. I went because I'm a citizen of this country." A strange, cloudy, guilty look came over his eyes. "There were three others from Yakutat in that war." I think he meant Indians.

"One killed himself. Got it mixed up in his head. He fought the Germans."

Abruptly changing the subject, he said, "My son is here, going to work up on the North Slope—over there." He pointed at the tall Indian. The young boy had long since left. The son was standing with his legs apart and his shirt buttons open. He had a goatee and medium-long hair. He was nicely dressed for someone from Yakutat. His girlfriend's tiny feet were between his, with her head lying against his chest and her hands in his back pockets pressed against his buttocks. He rubbed his hand lightly along her back.

At the word from the old man, he broke away, eyes bright with anticipation, but then he caught himself in half stride when the old man stopped to ask me my name, realizing that he was going through a parody for a drunken old fool. I told the son my name and shook his hand.

It was a classic performance by a broken, working-class father. I caught the look in the son's eyes; they were sympathetic. He was probably used to it. I half-winked, hoping he knew I understood, having grown up in a neighborhood of millworkers myself in the mountains of North Carolina. He quickly went back to his girlfriend.

Later, I was playing pool by myself at the Lodge. A good-looking, wide-shouldered black man was playing, too. "You make me feel brand new," came the words from the speaker. He had selected a string of blues and soul music to play on the jukebox.

We crooned to ourselves with the music as we ran the balls smartly around the tables. Click! Click! Clunk-clunk! I felt a kinship with him, though we never spoke at all.

On the way back to camp, I read a letter I had received from my mother back in North Carolina.

She wrote, "Just wanted to write. I get to wanting to see you so bad. Just remember you promised to bring the girls to see us. We mailed Suzanne's birthday gift, so she would get it before she left on her vacation . . . Your dad has a real good garden but I don't know when the tomatoes will be ripe . . . Glen's son, Mike, finished high school this year and is leaving for the navy on Wednesday. Your Dad is working on his bees and garden about all the time. The girls sent him a jacket for Father's Day. Be sure and don't go off again without seeing us, and bring some pictures of that beautiful country you said you was making. Take care of yourself and may God keep you safe."

Fly Away Tomorrow

Field Day

The seal winked
at me from the sea—
The mountains celebrated sunshine
in their heights—
Above
the low mist.

Clear sun on the
peaks' razor-sharp edges—
St. Elias' pyramid
only half seen—
through the layering.

Walking along
the steep beach—
I thought about
home—
with cold
on my hands.

My mind wanted
to express thoughts—
Unthought feelings
but felt dearly—
As we made a fast
pace to the end.

Billy Whiskers

In the quietness of our house in the wilderness at Cape Yakataga, the four of us talked from time to time and became close. Toward the end of what was another bad weather day I started talking with Gayle. I found her to be an intelligent and sensitive woman from a broken marriage, which you find so often in Alaska.

While we were surveying the profiles, Gayle would walk along the beach and sometimes just sit on a log looking at the surf. I asked her what she was thinking about during those times.

"Good thoughts. Just getting my head straight. I love it out here. It's been a good—a beautiful time for me."

That night I wrote in my diary, "Away from the slow euphoria of time abandoned, swiftly the deadline marches me back toward a more real world, as I drift knowing it's too short, wanting to hold on. Thinking already about getting away again.

"Drifting on the current soon to meet the chaos of the people sea with its complexities overriding simple work and sleep and eating and talk and horseshoes and interviews and touching and writing and reading and thinking and dreaming, alone yet together, peaceful."

Wednesday, 25 June 1975—
Cape Yakataga, Alaska

It was a beautiful clear morning! We had a 9:07 A.M. low tide, and we did five stations. The mountains stood out sharply behind the beach, and I took lots of pictures. At one station, we were eaten alive by mosquitoes. Our sampling had taken us well up into Yakutat Bay.

Friday was also a clear day, and we finished our stations at the head of the Bay. But Saturday the weather was bad again, and we were weathered in at Cape Yakataga. That night we went to a party at Minnie Eggerbrauten's place. Several geologists who worked for the state of Alaska, making an inventory of the oil prospects in the mountains around Cape Yakataga, were in attendance. Probably every major oil company in the western world had studied those rocks. But, for some reason, the state bureaucracy had decided to send a couple of former oil company employees and three or four fuzzy cheeked kids to check out the rocks. Why? Police work?

The leader of the geologists, Jerry Hardrock, was a mountain of a man about forty-five years old. He formerly worked for an oil company, but when faced with transfer back to the "lower forty-eight," he took the job with the state. We talked with him about clear-cutting and some state laws I had never heard of.

I had trouble communicating with Jerry Hardrock because he was hard of hearing, plus the fact that I mumble, so I had to yell a lot. He said he was losing his hearing because of his constant use of a helicopter to carry out the field work. I thought it was ironic that he came to Alaska to escape the bad things on the outside and was going deaf from helicopter noise.

Old man Reynolds, the weatherman, came in and gave us some of his blueberry wine, which was great, and we all became quite drunk over the next couple of hours.

This trip had been so different from the earlier trips in 1970 and 1971 with respect to the people I had met. Most of the people I met on the earlier trips were more concerned with exploiting than with preserving Alaska. But on this trip, I met Halibut John, Jerry Hardrock, and many other people who loved the land and were concerned about what was happening to it. Gayle and I talked about this as we rode in the back of a jeep along the dusty road back to the house. She was concerned about it, too.

The weather was bad again the next morning, a Sunday, as Chris and I made plans for the remaining work he and Janie would do after I had gone. I was supposed to catch the plane south to Juneau that evening. Though the weather was not favorable, at about 1:00 P.M. Gayle and I decided to take a chance and make a run for Yakutat, with an intermediate stop at the Yahtse River camp for me to say good-bye to the other members of the field crew.

Gayle and I were talking about the good times we had at the Cape and weren't paying enough attention as we took off and quartered out over the ocean. Distractions were something we couldn't afford then because the plane was bouncing around fiercely in the strong wind that was blowing at thirty to forty knots. Once out over the water, I could see the waves coming at a wide angle to the shore from the southeast. The tops of the mountains were covered with low clouds, and, in places, it was hard to see the gray outline of the shoreline. We skirted around the western edge of Icy Bay and finally went out over its iceberg-filled waters to Yahtse, where the wind was hardly blowing.

At Jon's camp, we exchanged information on what each field party had done, briskly going over our exaggerated accomplishments. I felt good about what we had finished. We had essentially completed all the work outlined in our proposal to NOAA. I knew that we had nearly completed the sampling program, had managed to relocate (and resurvey) most of our original survey stations, and had a series of excellent new photographs. I was looking forward to working with the data.

Soon we headed across Yakutat Bay to the airport with Jon on board. The fog was very low, and it was touch and go all the way. We arrived with not much time to spare. Jon would have to stay in Yakutat for the night because the weather was closing in.

As I was fooling around with my luggage and tickets, Gayle ordered lunch. When we finished the brief meal, she told Jon, "Well, I have to see the boss off."

We stood together in the terminal for a while discussing once again that special time we had spent together at the Cape. As soon as Jon joined us, I waved good-bye to them as I did a striptease in order to get through the homemade Yakutat security clearance system. Then I walked up into the 727, the last stage of leaving the field.

I stopped over for the night in Juneau, where I was met by my friend Cheryl. After dinner, we visited her friend LuAnn, who had a huge poster of Golda Meier above her dining table with a caption that read "but can she type." Both of these professional women were keen on politics and women's lib. I learned a lot from them.

Unlike South Carolina, the state I come from, which is fifty-first in everything,[22] Alaska has a lot of firsts: first in alcoholism, first in unemployment, first in suicides, first in abortion rates, first in birth rate, first in homicide, and so on. I obtained these data from LuAnn. I can't vouch for the accuracy of these firsts, but Alaska's population obviously had a few problems. I asked LuAnn why.

She said, "They sit in their one room in the long, dark winters paying exorbitant rent that they can't afford, trying to eat on food stamps, and they drive themselves crazy." LuAnn, a native Alaskan with three boys and a missing husband who did not pay child support, was incensed about what the logging and oil companies would do to Alaska.

On the way to the airport the next day, we stopped to observe the Mendenhall Glacier, and I was glad we came. The contrast of the blue ice with the dark green of the surrounding trees always gives this glacier a unique visual quality, but on this day, it seemed especially appealing with the clouds down against the surface of the ice between the mountains.

Once in my standard window seat, it felt good to be ready for another takeoff, because I reveled in the adventure of flying around in airplanes. And off we went down the rain-drenched plain at Juneau with green fir trees under the clouds at the mountain tops as the pilot made a sharp turn down the channel to avoid colliding with the mountain. The front of the elegant blue glacier, where we had stood an hour earlier, was just going out of view as the plane pushed up into the close gray of the clouds.

On the way to Sitka, I thought about Alaska, and I wrote the following in my diary:

So Alaska was a gem of rare and unequaled beauty, first discovered by white men at the birth of the nation by accident, by a British gentleman sea captain named Cook. Later it was squandered by the Russians, and then no one knew what to do with it until gold was found near the turn of the twentieth century when men were just learning to fly airplanes. Then it lay fallow, a land of busted dreams, except for a few traders, adventurers, homesteaders, and men in their magnificent flying machines, until the sixties, when people began to be concerned about the loss of natural beauty and to appreciate the genuine value of our natural resources. And then the grand bureaucracy of the early 1970s came to the rescue and:

1. Gave the land back to the Indians, who didn't believe in land ownership in the first place.

2. Gave the trees to the Japanese, who need the paper.

3. Hired some Texas cowboys to drill for the oil and to build the pipelines.

Predicted results:

1. The Indians will lose the land, and/or maybe a few will get rich.

2. The Japanese will have lots of paper and wood byproducts to sell to the U.S. at a huge profit.

3. The Texas oil men will get richer and richer.

4. The sad song of the fisherman will get sadder and sadder.

When landing in Sitka, you come in over the water. You don't see the runway until you hit the ground. I always thought we were crashing in the ocean.

"Welcome to historic Sitka by the sea."

The next day I walked the streets of Seattle. I saw an old man with one eye who could hardly walk, scouring through a garbage bin, walking from can to can with a shopping bag, and I made a mental note, comparing him with the recluses we had met in the wilderness.

An Indian in his twenties wearing a headband, drunk almost past walking, bumped into a black lady stumbling out of a bar. No room on the streets of Seattle to walk without bumping. In Cape Yakataga, where I had been, you could walk all day long without ever bumping into anybody, no matter how drunk you might be.

A little while later, I was walking up Sixth Street. The sun was setting. A jet plane passed with the sun reflecting off its wings. A few headlights flickered down the canyoned streets. Where were the trees, and the mist, and the mountains in this world made by men on these hills by the bay? In Seattle, the rain falls by the gallons uncounted, and the beat people gather under shelters of tin and cardboard. Seattle, city of concrete, city of lights.

As I walked farther along, the streets were nearly empty, even of derelicts. It was just me and the ghosts of the past walking together there.

I sent this description of Seattle, City of Concrete, to my friend Paul Hague, who sent me some notes that included the following words, which were spoken by Chief Seattle in a speech to Governor Isaac Stevens upon the signing of the Port Elliot Treaty in 1855. Seattle had just surrendered the lands upon which the city of Seattle now stands. In return, the Washington tribes were given a reservation.

But when the last Redman shall have become a myth among the Whitemen . . . When your children's children think themselves alone in the field, the store, upon the highway, or in the silence of the pathless woods, they will not be alone. In all the earth there is no place dedicated to solitude. At night when the streets of your cities are silent and you think them deserted, they will throng with the returning hosts that once filled them and still love this beautiful land. The Whitemen will never be alone.

Let him be just and deal kindly with my people, for the dead are not powerless. Dead—I say? There is no death. Only a change of worlds.

Friday, Independence Day, 4 July 1975—
Atlanta Airport

Sitting in the jet at the Atlanta airport late at night, waiting for the takeoff to Columbia, I thought about the job waiting for me back at the University of South Carolina. I had been head of the geology department for almost two years by then. I was asked to take the job only one year after arriving from UMass. I think I gave the faculty members and the dean the impression that I was halfway organized, a somewhat unusual characteristic around that free-wheeling, hustling department.

> You are holding
> in your hand
> the head of
> the Geology Department.

> _____
>
> Writing on wall above
> urinal in men's room of
> the geology department
> of the University of
> South Carolina, summer
> 1975

In the summer of 1973, I had informed Leita Jean that I had accepted the job while she was taking me to the airport for a trip to Iceland. She was planning to go back to school in the fall and major in geology. I forget exactly what she said, but I believe the words that hung in my mind and rung in my ears on the midnight Loftlieder special to Reykjavik were: "You stupid SOB!"

Head of an academic department means that you are sort of the boss. I figured it out just the other day: you are like a referee of a basketball game. You can call a foul now and then, but you can't throw anybody out of the game because they all have tenure. And the dean does the ultimate hiring and firing anyway, so you're operating from a doubtful base of power, at best.

Most faculty members shy away from the job because it is basically a thankless task, and besides, it eats into your research time. I took the job because I'm a supreme egotist, just in case you haven't figured that out yet, and I love being boss, no matter how dubious the role may be. There was also another motive. I was planning to go into business some day—we were already operating a wannabe educational film company, "Chugach Associates," on the side—and I felt the position of head of the geology department would give me the managerial experience that I lacked. And, in fact, I did learn a considerable amount about managing people during my first two years on the job.

When I got back to the university the next day, after a six-week absence, I found that my administrative assistant, Marion Reeves III, and my assistant chairman, John Carpenter, had things running more smoothly than when I left. Therefore, I just mostly signed the papers and kept my mouth shut for the rest of the summer, applying that old South Carolina management adage: "If it ain't broke, don't fix it!"

For the month of July, I worked on the Alaska and Kiawah Island data and reports, looking forward to taking a couple of weeks off in August to go fishing or something. On the last day of the month, I received a call from Joe Hartshorn, my former chairman at UMass, who had worked with us in Alaska during the 1969–1971 survey. He told me that the National Science Foundation (NSF) had called asking for an expert on New England and Alaska beaches to go down to the Strait of Magellan and check out an oil spill. The purpose of the trip would be to see if an analog study could be done to compare that area with New England and Alaska. Joe recommended me, a move that would change the course of my career forever. Later, the NSF representative called, and we made tentative plans for me to make a two-week trip to Chile in August, supplanting my previously planned fishing trip to the mountains of North Carolina.

THE END OF

THE EARTH

Crazy Norma

Sunday, 10 August 1975 —
Santiago, Chile

We got off the plane and walked out into the intoxicating clear, light air of central Chile. What a contrast with the hot, muggy smog we had left behind in Columbia.

I was traveling with Erich Gundlach, a new graduate student who had just arrived at the University of South Carolina after finishing a tour of duty in the Peace Corps in Chile. He came to USC-East with high hopes of raising money to go back to Chile to do a dissertation on some aspect of the coastal geomorphology there. The fortuitous circumstance of the phone call from the NSF asking me to do a project in Chile five days before his arrival in Columbia must have meant that somebody up there liked old Erich. Needless to say, his excellent skill with the Spanish language and previous knowledge of the country were invaluable assets on this trip.

As we rode the bus from the airport, I saw a sign (see the figure below), and it occurred to me that I had gone from one end of the Pacific coast mountain belt of the Americas to the other within the span of two months. And, as we headed toward Santiago, I looked up above the low haze that covered the city and saw the snow-covered tops of the Andes towering above us, just as Mt. St. Elias and Mt. Fairweather had done as we stood on the beaches in south-central Alaska.

In Santiago, we caught another bus to the coast to visit Erich's old stomping grounds at Vina del Mar, which is a wealthy suburb of Chile's largest port city, Valparaiso. As we came down over the hill into Valparaiso, I saw dozens of small boys flying kites over the steep slopes that slant down into the Pacific Ocean. The kites hung high in the clear sky over the closely clustered, thin-walled boxes that the local residents used for houses.

First, we made arrangements to spend the night in Erich's former residence, a luxury apartment in a French Riviera–like setting on the side of a hill in Vina. Shortly thereafter, we were walking along the beach when

FIGURE 3: *Road sign near airport in Santiago, Chile.*

we heard someone calling Erich's name. It turned out to be Patti Rojas, an old friend of Erich's, who at the time was hanging on the arm of one Richard from Mississippi, a marine biologist and another Peace Corps volunteer. Richard was thirty-seven on that very day. He was supposed to teach the Chileans something about their crustaceans, but I was puzzled as to how he was going to do it, since he couldn't seem to speak a word of Spanish.

We ate dinner with them, drank some *pisco sours*, the national drink of Chile, and decided to go dancing later. Patti said she would arrange a date for me with one of her aunts. We went to Patti's family apartment, in one of the high-rise apartment buildings overlooking the ocean, where she introduced us to the rest of her clan, including the ruler, Granny.

My date for the evening, Aunt Norma, was a real surprise, about twenty-eight and very pretty. They called her "Crazy Norma," and Erich was a little worried about how she would react to me. Keep in mind that Erich and I had known each other only about a week. They said Norma didn't usually like *Norte Americanos,* but she was lots of fun, showing me her paintings with *mucho* grunting and pointing. She spoke very little English, and I spoke about three words of Spanish.

While she was showing me her paintings, we listened to music. Suddenly, she got up and did a marvelous Latin dance. She did that sensational dance in the living room of the high rise overlooking the *Oceano Pacifico*, while the rest of us sat in stunned silence, even Granny who had already said that you could never predict what Norma would do.

After that rousing start, we went to the Topsy-Turvy, a discotheque on top of the hill above Vina. We slid down a fireman's pole to get to the dance floor, where we drank more *pisco sours*. Norma and I danced a lot, hugged a lot, and kissed a lot. We had a fabulous time, a night I will always remember. Unfortunately, we had to leave after a couple of hours because of the 1:00 A.M. curfew law. It had not been long since President

Allende was shot, and the country was under martial law, which meant everybody had to be off the streets by 1:00 A.M. They were serious about this—they would shoot you!

Outside on the hilltop, we were treated to a spectacular view. You could see the lights of Valparaiso sloping down, down toward the ocean for what appeared to be miles to the south. When we dropped off Patti and Norma at the apartment, we said we would come back by Vina on the way home. Crazy Norma wrote a note and handed it to me as I left, which said: "I will think of you in all moment (night and day). Since I meet you *y* love you."

The *Metula*

Monday, 11 August 1975 —
Vina del Mar, Chile

Next morning at 10:00 A.M., we were back at the airport in Santiago, sitting in a Lan Chile airlines Boeing 727 waiting to take off for Puerto Montt and Punta Arenas. The plane was crowded with geriatrics from South Africa on a low-cost "whisk" around South America. Somehow the crisp South African accents didn't fit into the scene. After a long delay, we were off into a heavy fog. As we broke out through the fog, I could see the high, snow-covered peaks of the Andes that would parallel our flight south.

We came down through heavy clouds at Puerto Montt over a flat landscape with patches of green fields and forests that looked a great deal like the countryside around the airport at Halifax, Nova Scotia. The 727 rolled to a stop in front of a group of buildings painted in camouflage colors. I disembarked to take a walk and catch a breath of the cool, forty-degree air. As I started away from the plane, I turned to take a picture of it framed against a background of the low-hanging gray clouds. When I snapped the picture, a military guard walked up and pointed his automatic rifle at me, indicating that maybe that wasn't such a good thing to do. I didn't take any more pictures.

Between Puerto Montt and Punta Arenas, we flew mostly inland, over the dry mountains of southwestern Argentina. We flew over the Strait of Magellan and landed at Punta Arenas, and my first impression was how bleak the place was. It reminded me a lot of Iceland. At first glance, I couldn't see any similarities with either New England or Alaska.

We were met at the airport by William Texara, another Peace Corps biologist who was stationed at the Instituto de la Patagonia in Punta Arenas. On the ground, Patagonia still seemed bleak but a strangely enchanting place. As we approached town, driving along the waterfront, I could see in the distance, far down the sound to the west, the end of the Andes snow white against the blue sky framed like a natural skyline behind the city, making one last brilliant show before plunging off under the Pacific Ocean to the south. We took a room in Hotel Turismo, which was located by the central square of Punta Arenas, a charming city of sixty thousand that sits on the north shore of the Strait of Magellan, at the end of the earth.

Then we went to the Instituto de la Patagonia and met the people there, including the *jefe*, Sr. Mateo Martenec, who opened the lab to us. They were a group of young biologists, geographers, and historians, a curious mix. They joked with us about the oil spill. We weren't exactly the first foreigners on the scene. The spill was a year old, and because of the political unrest in the country, I assume, no attempt had been made to clean it up. Thus, the spill site was essentially a natural laboratory where one could study how effectively natural processes could clean up an oil spill without man's interference. As you will see, the results were mixed.

Enrique Zamora, a young geographer of about thirty, was assigned to work with us. We asked for as little as possible, but they were very helpful, securing a car and driver for us, and we were glad to have Enrique's help. The instituto's secretary, Rosee Reyes Scott, who spoke English well, was also assigned to accompany us in the field.

On this trip, we were destined to spend eight days in the spill area, during which time we took over two thousand pictures, ran a few beach profiles, and collected several armloads of samples of beach sediments and oil. After we finished, we wrote a report for the NSF. Some of the choicest of the golden words of wisdom that appeared in that report follow:

On 9 August 1974, the Royal Dutch Shell super tanker VLCC *Metula* (206,000 dead weight tons; 328 m long) ran aground just west of the First Narrows of the Strait of Magellan. Forty-six days later it was refloated, but not before 51,000 tons of Saudi Arabian crude oil and 2000 tons of Bunker C fuel oil were released (Hann, 1975).[23] Only the *Torrey Canyon* disaster in Great Britain released a greater quantity of oil into the ocean, 117,000 tons according to Smith (1968).[24]

A preliminary assessment made shortly after the spill (Hann, 1974) estimated that 40,000 tons of oil were deposited along 75 miles of shoreline, with the Tierra del Fuego side receiving the major amounts. The quantity of oil on the beaches varied greatly along the coast, presumably a function of the wind, wave and tidal conditions, and land configuration. Biological impact was reportedly severe, affecting 600–2000 sea birds, intertidal mussel beds, marsh life, and nekton (Hann 1974; 1975).[25] However, a complete assessment of the biological impacts was never made. Biologists at the Instituto de la Patagonia are continuing study on the spill with minimal financial support.

A reconnaissance was made of the *Metula* spill area in August, 1975, to determine coastal morphological similarities between the affected area and future potential spill sites in New England and Alaska. Oil was still very much in evidence in the intertidal portions of the shoreline. The gravel sediments of the high beach face and the low-tide terrace were cemented by mousse-oil mixtures into layers several centimeters thick. A layer of mousse (water-in-oil emulsion) up to 1 cm thick remained on extensive areas of the tidal flats. In places, a dense blue-green algal mat was growing on the surface of the oil.

The Strait of Magellan dissects major Pleistocene glacial deposits, including terminal and ground moraines and terraced outwash plains. Inside the First Narrows, the sediment type and beach morphology show a striking resemblance to the Pleistocene glaciated shoreline of southern New England and to the present glacial shoreline of the south coast of Alaska. Tidal and wave conditions inside the narrows are also analogous to southern New England. On the Atlantic side of the First Narrows, the tidal range increases abruptly to approximately 10 m at spring tide. At low tide, tidal flats over 8 km in width are exposed. These tidal flats are very similar in appearance to tidal flats of Cook Inlet and in the Copper River delta region of southern Alaska.

The Patagonian Desert

We had an appointment to fly at 9 : 30 A.M. the next morning. Enrique took us to the wrong airport, where our papers were looked over earnestly and at length by a number of army guards. It was a strange feeling to be surrounded by those submachine guns, something like surfacing somewhere in the middle of a grade B World War II movie.

We finally made it to the right airport and were up and away over the most spectacular drumlins in the whole universe. *Drumlins* are streamlined hills of soil and sediment left behind by the glaciers that covered this area during the Pleistocene glaciation. We flew along the north shore of the Strait looking for traces of the oil and making mental comparisons with the other end of the earth. Clear impressions of the old glaciations were everywhere, left behind by the ice sheets that have come down out of the Andes from time to time. The last one left there about ten thousand years ago. Its terminal moraine and braided outwash stream patterns, being only partly obscured by the semi-arid steppe vegetation, could be easily discerned from the air. In fact, the present course of the Strait follows along the older valleys gouged out by the ice. We took pictures of these features plus a variety of coastal land forms.

When we got to the First Narrows, the pilot said we had to go back because he was running out of gas. I don't know how we made it back. I felt stupid for not checking the fuel gauges before we left. Erich yelled a lot in Spanish, which I didn't understand. When we got back to Punta Arenas, the pilot crash-landed in a strong crosswind, possibly the worst landing I had ever experienced.

We spent the next day talking with people at the ENAP (the national oil company of Chile) offices, looking at their aerial photographs of the spill area, and getting organized to go into the field the next day. Our field crew consisted of Enrique and Rosee from the *instituto* and Alfredo the driver, with his trusty Volkswagen bus.

So we drove out of Punta Arenas in the dark at 6:30 A.M. on

FIGURE 4: *Sketch of Darwin's rhea.*

15 August 1975. We watched the sun rise over Tierra del Fuego as we drove through the ghost-like drumlins we had seen from the air. All the day, which was *muy luminoso,* we drove through sheep-covered glacial hills, coated brown by heavy stands of grasses. We saw an abundance of strange wildlife, including numerous huge ostrich-like birds (rheas)[26] feeding among the sheep, gigantic jackrabbits, foxes, pink flamingos, and wild geese. The landscape reminded me of eastern Wyoming, eastern Colorado, or West Texas. The bright light of the rising sun shone at a low angle, and the white sheep dotted the brown everywhere.

Maybe I should inform the reader that up to this point, I had never been to an oil spill. In fact, I did not have a clue on how to go about studying one. But I knew we had to make systematic observations, so I employed a technique called the "zonal study" that we developed in Alaska in the summer of 1970. In a zonal study, you lay out a systematic grid of sampling and beach profiling stations designed to give a representative account of all the different shoreline types occurring within the study area.

When we walked out on the beach at Station 2, I was amazed at the striking similarity with the beaches in New England. Except for the sheep and the *coeron* grasses in the background, we could easily have been at Winthrop Beach, Massachusetts, or at any of a dozen places along the south shore of Cape Cod.

At Station 4, we walked about two miles across the steppe to the beach. On the way, I took a picture of a sheepherder in his red shirt and black hat and with his wind-burned cheeks posing outside his bleak box on the steppe with two pups standing guard. On the way back we became lost, so we split up, walking in opposite directions along the road. I talked with Rosee as we walked. She told me she had studied journalism for four years in college, but now, as secretary at the *instituto,* she was "unhappy with herself" because "she wasn't doing anything." Somewhere I had heard that story before. But she loved Punta Arenas and wanted to stay there with her family. The other two finally located Alfredo the driver.

Toward the end of the day, we drove along the high rim of the cliff on the north shore of the First Narrows. For a while, I watched the sun set on my left, and then I looked down over the cliff to the blue water at the end of the earth, stained black around the edges by the one-year-old oil spill.

Then I looked ahead at the lights of the burning gas fires that marked the position of the oil wells on both sides of the Strait and asked myself, "Why is there oil at both ends of the world, and especially here under these Cape Cod hills covered with Wyoming grasses?"

We spent the night in ENAP's camp in Posesion, a cluster of houses in a valley between two glacial moraines. We ate in a dining hall with Chevrolet trucks parked in neat rows out front. The oil men all look the same, whether it's West Texas, the Alaskan tundra, or Patagonia, wind blown and tough, earning their pay. We slept in their bunks. The army was there also, driving Toyota trucks and standing in lines, dressed in green with short hair and mustaches and carrying machine guns.

The Strait of Magellan

Saturday, 16 August 1975 —
Posesion, Patagonia

As we left Posesion the next morning, the black clouds in the sky reflected darkly on the yellow-brown hills. We went first to Cabo Dungeness, a huge accumulation of gravel beach ridges that guards the north entrance to the Strait. It is a classic cuspate foreland, a triangular-shaped pile of beach sediments that builds out into the water where large waves coming from two opposing directions meet along a shoreline. This cuspate foreland is undoubtedly named after Dungeness in the English Channel, a feature formed by exactly the same type of coastal processes. The English sailors that named this cape during an early passage through the Strait probably didn't know why they looked so much the same.

Cabo Dungeness is flanked by a high scarp on its north side. We stood on the top of it and looked down on the majestic foreland, which was dotted with fires flaring gas at the ENAP wells.

We drove down the scarp and over the gravel ridges out to the end near an old and rusting lighthouse, where we surveyed a station. With my shovel, I probed into a gravel layer in the beach, which had the consistency of bubble gum. That artificial rock layer, which was held together by the *Metula* oil, had a dead cormorant's wing cemented in it.

I walked away from the beach, with oil on my knees and hands and all over me, careful not to step on the dead birds, and I thought about Lake Erie and the dead fish that I had walked on there. And as I used the gas siphoned by Alfredo onto the cloth that Rosee gave me to wipe my hands clean, I swore I would tell it and tell it and tell it!

I continued raving as we drove back out the road. "Must we have Lake Eries and great Patagonian oil spills until there are no more clean beaches for our children to play on?" Rosee seemed upset by my temper tantrum, so she finally said so few *personas* really cared. I could only get angrier. At the next station we saw tar blown up the scarp and a pond filled with ENAP drippings.

In the early afternoon, the sun came out as we worked our way back around the north shore toward the First Narrows. At Station 6, I saw oil coatings on the boulders, but the mussels were clean. "Maybe they eat it," I thought. (Later, I figured out that oil sticks to dry rock surfaces but not wet mussels.)

In the late afternoon, we climbed down a high cliff with a gigantic scallop where a massive piece of the cliff had slumped into the ocean. I marveled again at how much it all looked like Cape Cod. Then I thought about the glaciers and how, as they retreated, the people had moved in to eventually find the oil, as we will someday do in Antarctica, if the lawyers can talk long enough to get tired and finally sign the papers.

On the way out, we stopped again to check out the inspirational view from the top of the cliff. I stood there, trying to picture in my mind what the landscape in front of me must have looked like when it was covered by glaciers during the last ice age. It had only been a few weeks since I stood at the terminus of the Malaspina Glacier, wondering about the terrain under that massive ice sheet.

At the close of the day, as we drove to the ferry landing at the First Narrows in order to cross over to the Tierra del Fuego side of the Strait, we saw two gray Patagonian foxes, more pink flamingos, geese by the hundreds, and rheas by the dozens.

Alfredo nosed the Volkswagen onto the small ferry, making a loud thump when the back bumper struck the side of the dock as the boat sank under our weight. As the ferry pulled away from the dock, I walked up to the bow of the *barco* and looked back to the shore of Patagonia in the light of the sunset that lingered in the near darkness, and then to the other side to the fires of Tierra del Fuego. And thought about where I was and tried to remember how far away it was to Yakutat, Alaska, the place where I had first heard the sad song of the fisherman. No, not entirely, I was mostly just feeling the wind and the cool of the evening on a fifteen-minute crossing of the Strait of Magellan.

The Great Patagonian Oil Spill

They burned
the fires
here and there
on the plain,
and the sun
hung low

over a West
Texas–like terrain.
Tame sheep and wild geese
and Patagonian foxes
played among the
ice made hills
around the houses
built like boxes
in the Great
Patagonian Oil Spill.

These fires that
burn eternal
mark the channel
for the shipping
on the
Patagonian desert
where the tides
go through a ripping
at the sides
of the tight channel
of *Estrecho
de la Magallanes*
in the Great
Patagonian Oil Spill.

The hills they
showed us beauty
in the grasses
luminoso,
but their sides
were sometimes plastered
with black oil
that had been blown there
by the winds
so strong in winter.
And the ponds
were full of drippings
that oozed out
of the drillings
in the Great
Patagonian Oil Spill.

I saw the oil
cemented cormorant's wing
on the beaches
in the gravel,
and thought
about Lake Erie
and other places
where I travel.
So with anger
total full and
completely overrun,
I suddenly concluded
we had just
run out of fun
in the Great
Patagonian Oil Spill.

Billy Whiskers

Field Days—Tierra Del Fuego

Sunday, 7 August 1975—
Sombrero, Tierra del Fuego

We spent the previous night in another ENAP camp called Sombrero, which is located on the Tierra del Fuego side of the Strait. We stayed in an elegant house in that oil company oasis that looked a lot like an American suburban housing development, with abundant electricity, a gymnasium, a church, and *mucho* Chevrolet trucks.

This particular morning went slowly as we spent a lot of time pushing Alfredo's bus out of the muddy ruts in the roads. It was a rainy, bleak, cold day. We talked little as we moved like automatons performing a religious ritual by carrying two sticks and a tape in jerky motions across the surfaces of the gravel.

About 3:30 P.M., we drove onto Punta Catalina, a truly spectacular cuspate foreland that guards the south entrance of the Strait, like a mirror image of Cabo Dungeness on the north side. The strong westerly wind is eroding some of the older, now elevated dune lines, creating scalloped ghost-like forms that look a lot like some of the features seen in photographs of Mars. We did two stations and then drove out to the end of the foreland under a clearing sky, looking into the sinking sun. The low gravel beach ridges stood out in bold relief as shadows filled the shallow valleys between them. At the end of the foreland, we stood facing the open Atlantic Ocean; the sun dazzled on the clean rounded pebbles on the beach.

After running a profile and sampling the gravel, we drove back across the perfectly lit plain. A single horse, standing on the rim of the high scarp on the south side of the foreland, watched the multicolored sunset in quiet solitude. I was writing, trying to find words to describe the overwhelmingly resplendent sky.

Rosee asked, "Erich, what is the *inglés* for *inhóspito?*"

"*El mismo* (the same)," he replied.

Then she looked at me and said, "For a land so inhospitable, we must have a sky like this." Her eyes sparkled when she said it, as if she knew she had answered my unspoken question.

It was a perfect ending for a bad weather Sunday. I wasn't the least bit depressed. In those days, I usually went into some kind of a deep depression on Sundays. My friend Sally said it was because I didn't go to Sunday school any more.

On the way back to Sombrero, we drove across the tops of the till hills looking northwest at the fading light of the sky. One very bright star hovered over the Strait. Spanish songs were playing on Alfredo's radio. Occasionally an English one could be heard—"Oh yes, I'm the great pretender . . ."

As we approached one of the burning wells, smoke from the fire stretched all the way across the faint yellow horizon at a thirty-degree angle, blown by the westerly winds. The star was just above it. The air pollution seemed puny indeed in that vast barren land of Tierra del Fuego. I watched the reflections of the dying light on the ice-covered ponds. We were each lost in our own thoughts, except for Erich who slept and Alfredo who fought the ruts.

A perfect, sharp clear day dawned on 18 August, which was excellent timing because we planned to fly to take pictures on that day. We drove down to the one-room air terminal at Sombrero and waited for the plane in surroundings that reminded me of the air terminal in Cordova.

It was a perfect flying day, as we marveled at the five-mile-wide tidal flats and the spectacular cuspate forelands at the entrance to the Strait from a variety of altitudes in our chartered Cessna 172. The oil was everywhere. We estimated that it was spread over more than 250 miles of the shoreline of the Strait, considerably more than Roy Hann's original estimate of 75 miles. Back at the airport, we told Enrique and Rosee good-bye, and they took the plane back to Punta Arenas.

Late in the day, we surveyed another typical Cape Cod beach as the sun dropped behind the Andes in the far distance. The place was called Punta Remo; a thick layer of oil was at the high-tide line.

The people at the *instituto* recommended that we spend that night at Hotel Bahia Felipe, which we located on the map near our westernmost sampling station. So we headed for our sleeping quarters, bumping along the ruts in the dusk of the fading light.

As we approached the hotel, I was awakened by a surprised exclamation from Alfredo and looked up to see a yellow-and-white Cessna 172, a twin of the one we had been flying in, sitting crossways in the road. Apparently, the pilot had landed it in the road after engine failure. It was a strange specter in the dark. Even stranger was Hotel Bahia Felipe, which was nonexistent. The headlights of Alfredo's bus, which were supposed to be centered on the hotel, pierced out into space over the marsh. The hotel had been torn down.

With no place to sleep and anticipating another hour or so of bouncing in the ruts, we wearily climbed back into the bus and headed for Porvenir, a small town located directly across from Punta Arenas on the south side of the Strait. We arrived about 8:00 P.M. and obtained a room for three dollars in an ancient hotel at an intersection of two dirt roads.

After dinner, we decided to check out one of the night spots of Porvenir, an activity Alfredo had been recommending enthusiastically. Three or four bars were located on one of the side streets. We chose the one with the red light above the door. That was our first mistake.

It was a relatively small bar, but economically laid out. Up on an elevated bandstand against the wall, a four-man band played awful music. And the posters! Every wall was covered from top to bottom with posters of nude ladies. The one I liked best was an eye chart superimposed over a nude midsection.

We took a table and ordered *piscolas, pisco* wine added to Coca Cola. All the women sat in the opposite corner by the bar. Alfredo asked the prettiest one, whose name was Veronica, to dance. Alfredo brought her over to talk with us, which she did off and on for the rest of the evening. Alfredo spent most of his time with one of the girls named Rosita.

As the place gradually filled, a whole lot of dancing was going on. One well-dressed fellow, maybe fifty with a pot belly, a huge beak nose, and long thin legs, danced skillfully around the floor, lifting his feet straight up behind him at ninety degrees to his upper legs, which appeared to hardly move at all. His feet flipped up and down to the beat of the horrible noise that the band was making. By the time he started doing his thing, I was pretty drunk, so I amused myself by playing eye games. I would squint my eyes, and the old man disappeared, changing into one of the giant rhea birds we had seen on the hillsides, dancing away with the blue-skirted Veronica. I would squint again, the bird would go away, and the Fred Astaire of Tierra del Fuego would reappear. Sometimes you run out of things to do when you're waiting for closing time.

Another dancer, a young guy about twenty-five with a face like Jack Palance's and with mud up to his knees, wasn't doing too well. He was turned down three times for one of the numbers. Erich allowed that we were watching the eighth wonder of the world, a man who couldn't make out in a house of ill repute. After that pronouncement, Erich left, leaving Alfredo and me alone with Veronica and Rosita.

A little while later, two rough-looking cops stood just inside the door. The place was deathly quiet for a long time, even after they left. And about 12:30, somebody opened the door, and a blizzard-like, frigid wind blasted across our table. I closed the door, but Rosita opened it again, laughing at me. I guess that must have been the signal for closing time.

Later, Alfredo and I staggered down the frozen street to the corner hotel and our refrigerated room. The room contained two beds pushed side by side, barely fitting into the small space. Alfredo and I kept our clothes on and rolled up in the covers like cocoons, hoping we wouldn't freeze to death.

At 7:15 A.M., Erich's clock went off, and he turned on the light. Alfredo and I sat up, trying to untangle ourselves from the covers. I looked over at him and he looked back, with a sheepish grin on his face. I was numb from head to toe.

"*Malo noche!* (bad night)" I said.

"*Si! Si! Muy malo noche!*"

"We're supposed to have breakfast at 7:30." Erich said.

We walked down the dark hall and into the kitchen where an old man about eighty was poking around inside a wood stove. I hovered over the stove for a while, the only heat in the whole building—and perhaps in all of Porvenir.

The old man shuffled slowly about his chores. He had ghastly looking eyes, bloodshot under open lower lids, the eyes of death. He put a plate with a few crusts of stale bread on the table along with three cups of coffee that you could walk on. When Erich asked him if he had any butter, he politely informed us that we weren't staying at the Hilton.

Out on the streets, Porvenir looked more bleak than ever, with the stark gray walls and steep streets framed against the gray of the sky. We loaded into the bus and headed back for Hotel Bahia Felipe. On the flat hilltop, we could see in good light the skeleton of the hotel scattered over the ground and the Cessna 172 sitting catty-cornered across the road.

On the shore by the destroyed hotel, we saw abundant oil again, especially on the mud flats behind the beach. It had seeped through the sand and accumulated at the impermeable layers at depth. When I dug into the banks of the tidal channels, fresh oil floated on the surface of the water out toward the sea.

Our next stop was the ferry landing at Punta Espora on the north shore of the First Narrows, where we walked on the worst oil of all. A zone of the intertidal gravel at least ten meters wide was saturated with oil and looked considerably like a recently paved runway for a backwoods airport. Remember, this was one year after the spill. And as we walked back to the bus, with the stench of the oil in our nostrils, I could picture that mess on the south shore of Massachusetts or on the mud flats of the Parker River Estuary on the North Shore, with little kids playing on it smeared with oil from head to foot.

We rode the ferry across the Strait, our field work finished. I stood at the rail the whole way, enjoying the sun and the wind. I had a two-hundred-millimeter lens on my camera, taking pictures of a gull that sailed along with the boat, framed in front of a perfectly blue sky. Suddenly, two dolphins surfaced behind the boat. They were white all over with black fins and snouts. Their colors were stunning against the green water.

Homecoming

Tuesday, 19 August 1975—
Punta Arenas, Chile

We arrived back at the *instituto* to a hero's welcome. Enrique was bubbling with enthusiasm. The director, Señor Martenec, shook my hand and babbled. I couldn't understand what he said, but I heard the words *muy rapido* and *muy importante* several times. He said he thought our work was important and that he hoped we would come back for more cooperative ventures with the *instituto*. I told him I would write a positive report, with respect to the comparative study, and that I hoped we would be back.

Rosee gave us both a big hug and a present with a note that said: "Miles and Erich—Thank you for those incredible days just past, especially the warm thoughts and the friendship . . . It was all *fantástico*, like the sky and the grasses of Patagonia. Rosee."

We also got a hero's welcome back at Hotel Turismo. I didn't know conquering the steppe was worthy of such praise. When we returned to Columbia, South Carolina, and the university, there were lots of papers to sign and problems to solve, but no hero's welcome. Such is life.

We had dinner with William Texara and his wife, Jean, who had served as a field assistant on the earlier studies of the spill. After dinner, she showed some slides she had made during the early days of the spill of penguins and cormorants running on the beach, brandishing their newly acquired oil overcoats. She also showed some eight-millimeter movies of the same thing. That was a heavy trip; I was brooding the whole time. It was worse than I had imagined.

Jean and William told us that the people at the *instituto* felt a little victimized—if not, indeed, used—by the politics of the spill. They described the visitors to the *instituto*. They found the PR man from Shell and the bureaucrat administrators from NOAA, my benefactors in Alaska, to be the most obnoxious. They liked Roy Hann, about the only scientist who had actually looked at the spill. They showed us a report by a British consulting firm hired by Shell. It reported that all the oil that spilled out of the *Metula* had been carried into the Atlantic by the tide. I was forty years old at the time and had been a practicing scientist for about eighteen years, but that was my first introduction to a theme I would hear many times in the future on what I now call "Science, Paid Liars, and Videotapes."

On the way back to South Carolina, we stopped for the night in Lima, Peru, a city built by the Spaniards on a high, bone-dry marine terrace. The steep face of the terrace drops straight down into the blue water. The city had magnificent squares with large cathedrals and statues of dead heroes.

We hired a cab, drove to the beach, and looked at the conglomerates in the high terrace and at the groins built to save the road. A *groin* is a pile of large rocks (riprap) placed in a line perpendicular to shore to arrest sand transport and, hence, slow down beach erosion. A spectacular sea stack, an isolated rock island carved out of the rocks by the waves, stood guard a few yards offshore.

Later that night, we prowled around Lima's *cantinas*, stashed in among the warehouses and factories. We heard some Americans laughing at a table, and the women sat across the bar not acknowledging our stares. Why were we there?

Our cab driver, who had a degree in economics, talked about the border dispute with Chile and the rising inflation in his country. He said the Indians should all be put on reservations. He would soon be heading for Australia to find work.

At 1:00 A.M., as we were walking back to our hotel after failing in our attempted conquest of the Peruvian women, we saw two men fighting over a two-cent (U.S.) increase in bus fare. A crowd had gathered to watch. Then we walked on, the stench of the sewer in our nostrils.

A little farther, we admired the volcanic rocks from which had been built a marvelous religious edifice for the seventy-five percent Indian populace to worship in.

I was thinking, "Two and a half centuries after Juan Pizzaro found the Inca gold, Captain James Cook, English gentleman sea captain, found a land by accident that would bring another kind of gold rush one century later. Another century after that, Billy Whiskers, descendant of Captain Cook's relatives, flew into Anchorage in a B-747 (built in Seattle) to become involved in the Great Alaska Oil Rush, and then four months later into Lima on a B-707 (also built in Seattle) to be driven around town looking for *señoritas* by a descendant of Pizzaro out of work. Meanwhile, a few of the Indians in Yakutat are still skinning and eating seals the same way they did when Cook arrived, and if you go far enough back into the mountains of central Peru, you may find a few Inca Indians working in the terraced farms just like they did when Pizzarro put in his first appearance."

My train of thought was broken when I looked up at the following sign:

INCA   COLA

FIGURE 5: A road sign in Lima, Peru.

We walked on to the hotel. I was tired—dead—enough. The trip to my first oil spill was over.

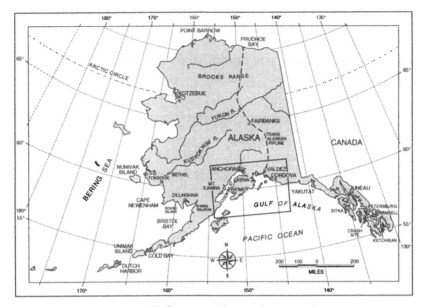

FIGURE 6: Location map, Alaska. Box indicates location of Figure 7.

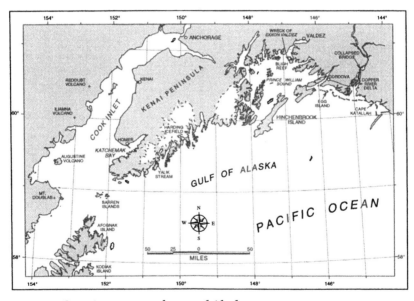

FIGURE 7: Location map, south-central Alaska.

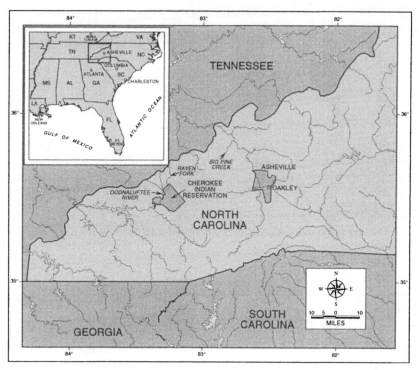

FIGURE 8: Location map, western North Carolina.

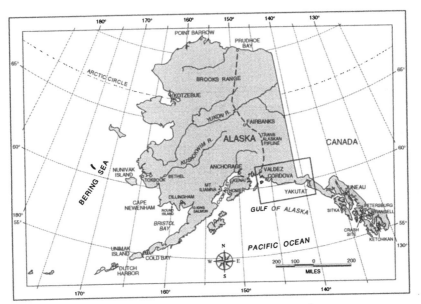

FIGURE 9: Location map, Alaska. Box indicates location of Figure 10.

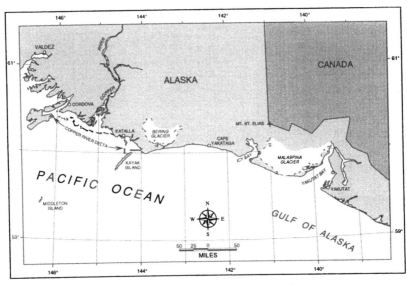

FIGURE 10: Location map, Cordova/Yakutat area, Alaska.

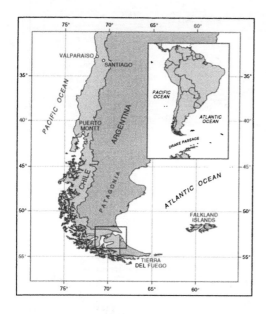

FIGURE 11: Location map, southern South America. Box indicates location of Figure 12.

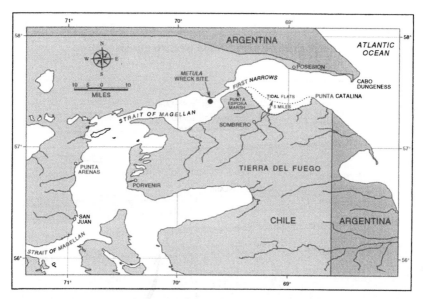

FIGURE 12: Location map, Strait of Magellan.

FIGURE 13: Location of three oil-spill sites in western Europe.

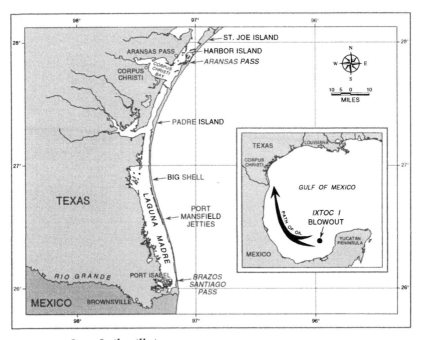

FIGURE 14: Ixtoc I oil-spill site.

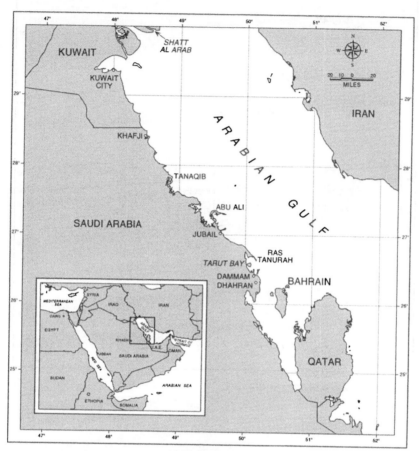

FIGURE 15: Gulf War oil-spill site.

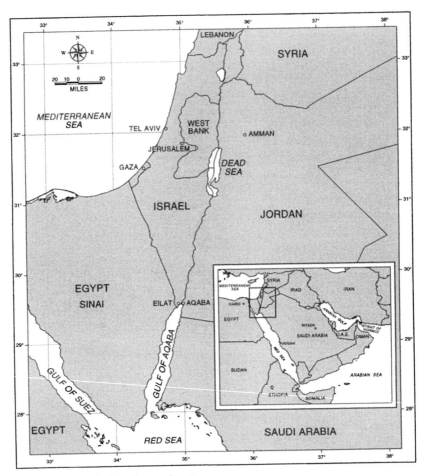

FIGURE 16: *Gulf of Aqaba study site.*

FIGURE 17: *Photograph from "Suzanne's Lament" (1970) of Jesus rock on Kayak Island, Alaska.*

FIGURE 18: Photograph from "Suzanne's Lament" (1970) of low-tide beach near Cape Yakataga, Alaska.

FIGURE 19: Photograph from "Suzanne's Lament" (1970) of a sea arch on an outer rocky shore west of Cordova, Alaska.

FIGURE 20: Jon's field camp at Yahtse River during the 1975 Alaska survey. Mt. St. Elias is in the background.

FIGURE 21: The gravel beach ridges at Punta Catalina—at the end of the earth (1975).

FIGURE 22: *Cristobal on the massive asphalt pavement at the First Narrows, Strait of Magellan, one and one-half years after the* Metula *spill (1976).*

FIGURE 23: *Field team arrives at Strait of Magellan, 1976. Left to right: Cristobal, B.J., Erich, Moon, Meelace, Ana, and Marcos.*

FIGURE 24: *The black tide of La Coruña, Spain, May 1976. Heavy mousse being deposited along the shoreline by the waves.*

FIGURE 25: *Bob Stein and Larry Ward of our team survey a profile across a heavily oiled beach near La Coruña, Spain, May 1976.*

FIGURE 26: *First beach landing on the eastern shore of Lower Cook Inlet during the Alaska surveys of 1976.*

FIGURE 27: Heavily oiled marsh at the Amoco Cadiz *oil spill, 1978.*

FIGURE 28: Cleaning an oiled seawall at the Amoco Cadiz *oil-spill site, March 1978.*

FIGURE 29: JM replies to a reporter's question while the cleanup team ponders what to do next at the Exxon Valdez *spill (June 1989).*

FIGURE 30: The cleanup of the Exxon Valdez *spill in full swing during the summer of 1989.*

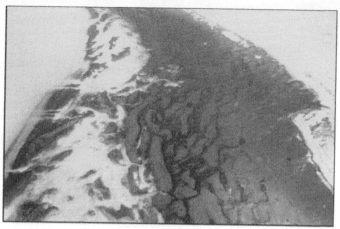

FIGURE 31: *Heavily oiled intertidal flat, several hundred yards wide, Gulf War spill. Photo taken by JM in May 1991.*

FIGURE 32: *The great field team at the Mother of All Spills in the Saudi desert. From left to right: Todd Montello, Abdul Halim Al Momen, and Ahmed Al Mansi.*

FIGURE 33: *Liquid oil in animal burrows (probably shrimp) in mud flat, one year after the Gulf War spill.*

PART 5

TEXAS
MEDICINE

Back to the End of the Earth

Monday, 20 October 1975 —
Columbia, South Carolina

On this date I was forty-one years old, but I felt like a new man because something different was happening. JM and I were seeing each other and would live together before another month had passed. I didn't know it for sure quite yet, but I thought that maybe I was no longer "in between dances."

I wrote a report for Ralph Perhac of the NSF on the *Metula* spill, and he was so impressed with the results that he requested a new proposal to do more detailed studies of the spill at the start of the following year. Therefore, on Thursday, 29 January 1976, our team was on the way back to the end of the earth for a much more detailed study of the *Metula* oil-spill site than the one Erich and I had done the previous August. We pondered how we were going to pay for the operation, considering the political instability and the worthlessness of the Chilean money. I'm sure Shell Oil Company would have done things differently, but I worked with what I had, namely the expert opinion of Erich who had lived there for two years through the chaos resulting from Allende's assassination. His solicited advice was "take cash" because everyone down there loved to be paid in U.S. dollars. We worked out a budget and concluded that we would need twenty-four thousand dollars to fund the entire field stay, which was scheduled to take six to eight weeks. This sum included the purchase of two vehicles and airplane rental.

For some reason, probably because I was extremely busy and preoccupied with my new relationship with JM, I didn't take a single journal note during that trip, so I'm writing some of this from memory. JM saved the eight long letters I wrote to her while I was there, and they have been a useful source of information.

The day before we left, I managed to talk the university comptroller into writing me a check for twenty-four thousand dollars, which I took to the local bank to cash the next morning. Getting that much cash took more persuasion than obtaining the check in the first place. Finally, the teller sent a uniformed guard downstairs to get the money, and he came back with a large paper bag stuffed with fifty- and hundred-dollar bills. Then the guard ushered me out to the street, the main drag in town. From there, I walked to the student health center, carrying the paper bag, to get the rest of my shots for the trip.

As the nurse was preparing my arm for the inoculations, I asked, "Ever see twenty four thousand dollars?" And then I opened up the bag for her to look inside. She jumped back, not knowing whether to proceed with the shots or call the cops.

Back in the office, we made last-minute preparations. I had already said good-bye to JM earlier that morning. My immortal parting words were "Well, I hope you're still here when I get back!" That's another one of those things that she has never let me forget over the past twenty-one years.

Just as we were getting ready to leave the office for the airport, my administrative assistant told me I had an urgent phone call from the bank. It was the teller who had given me the cash. He said, "Oh, Dr. Hayes, thank God I've found you."

"Well, actually, we are just leaving for the airport. What's up?" I responded.

"Could you please come by the bank on the way to the airport, it's urgent!"

"Sure, the bank is on the way, but we don't have much time."

When we pulled into the parking area at the bank, the teller and several armed guards were waiting for me. It seems that the teller had made a mistake. The paper bag actually contained twenty-nine thousand dollars, something I had not as yet discovered. The teller then retrieved his missing five thousand dollars but probably not his job. Such is life.

One of my assistants bought a camera bag for me to carry the cash in. It was a beauty, made of burnt-orange fake leather—very inconspicuous. Ha, ha. It held only eighteen thousand dollars, so before we got on the airplane, I distributed the rest of the dollars among my fellow travelers.

In Santiago, we bought two Argentine jeeps for two thousand dollars each and some other gear, and on Saturday morning, 31 January, we headed south on a twenty-five-hundred-kilometer drive, which would take us through Chile, over the Andes into Argentina, and then on through Patagonia into Punta Arenas.

Leaving Santiago, we drove down the four-lane super highway in the central valley of Chile, dodging the *pilegros* (holes in the road) along the way. It was a while before we escaped the smog of the city. The hills were pretty, with an arid brown coating of sparse vegetation. We all said the countryside looked like southern California. The air was light and exhilarating, also just like California.

We saw evidence of considerable poverty in the clustered huts on the flat areas outside the city. We stopped for lunch by a river in Taka. The river was a typical, arid-country braided stream and a little muddy, but people were swimming in it. One kid swam all the way across with a stroke a lot like JM's. She swam competitively in high school, and she was trying to improve my mediocre swimming stroke during our daily forays in the university pool. I already missed her.

The seven of us riding in the two jeeps made up the entire American contingency working on the project. Erich had made arrangements for a total of about twenty Chilean scientists to join us in the field at some time during the study. The first group was scheduled to meet us when we arrived in Punta Arenas.

The trip between Santiago and Punta Arenas took five days. We saw gorgeous country along the way, most notably the trout streams in the Chilean Andes. I plan to go back there some day to fish those streams.

Each day, we drove long into the night. We switched driving duty from time to time, and I can remember driving late one night in a rare rainstorm in Patagonia, which had turned the dirt roads into soft mush. I strained to see the swerving ruts through the mud splattered windshield and follow the two red dots of the rear lights of the jeep ahead that was just disappearing over a rise. Rain slashed across the headlight beams at a thirty-degree angle. Then we crossed the rise, and the jeep in front was again fighting its way out of a rut. We spent that night in a house of ill repute in a small village, the only sleeping quarters for hundreds of kilometers.

After a few drinks in the bar that night, Chris Ruby, shouted "drinks for everyone on me!" It cost him about three bucks, and the place was crowded, too.

On the way down, we decided that all the Americans should have Latin names. I was to be called "Meelace"; Chris Ruby was "Cristobal"; Erich, who was beginning his Ph.D. work on the spill, was just "Erich" because we couldn't think of a Latin equivalent; Anne Blount, who was working on an M.S. thesis on the oiled sediments, was "Ana"; Ian Fischer, working on an M.S. thesis on the tidal flats, also didn't have a Spanish equivalent, but his nickname was "Moon-man" so we just called him "Moon"; and Mark Cable, one of the field assistants in Alaska the previous summer who was assigned to help Moon, was "Marcos." In addition, Professor B. J. Kjerfve, the physical oceanographer in the geology department at USC-East of "dumb Swede" fame, was with us to study the tides in the Strait, but he would not be a part of our field team.

Once we got to Punta Arenas, it took a couple of days to make all the arrangements for our return to the field. Eight Chilean scientists met us to go out in the first wave. Saturday night, one week after we left Santiago, was our last night in Punta Arenas, and the whole group, which consisted of the seven Americans, all the Chileans going into the field, and all the Chileans who had helped get stuff together in Punta Arenas, a total of over twenty people, went out on the town. We went first to the best restaurant in Punta Arenas for a late Chilean dinner (dinner was always late). They seated all of us at one large table, with me at one end and Marcos at the other. As we waited for dinner, we started making toasts. I initiated the action by proposing a toast to the *Metula*, without whose sinking none of this would have been possible. Every one made a toast, with Marcos being the last in line.

It took some prodding, but Marcos finally stood up, raised his glass, and said, "To us, we are the best!" Then he sat down, looked around a bit sheepishly, and concluded, "I think."

After the dinner, we moved down the street to a small bar and continued the festivities. That is, until fifteen minutes before curfew time when we left the bar and ran down the street to Hotel Turismo.

Next morning, a scheduled meeting in my room took place at 8:30 A.M. to make final plans for the trip into the field, which would take place as soon as we finished the meeting.

As we were getting started, Natalia, a biologist from the Oceanographic Institute in Valpo, looked all around the room and finally asked me, "Where is it?" The "it" was the burnt-orange fake leather camera bag with its American cash, which the day before I had counted out to be exactly sixteen thousand smackeroos.

"Oh my God," I thought, "I left a sixteen-thousand-dollar tip in that bar!"

Erich, Natalia, and I raced down the street to the bar, only to find it closed with no living soul in sight. After all, it was nine o'clock on a Sunday morning. Natalia roused some neighbors, and they told her that the bar owner lived several blocks away. We went to his house, and he agreed to go back to his establishment with us. When we got there, with a big smile he reached behind the counter and produced the unopened burnt-orange fake leather camera bag. Needless to say, we thanked him profusely. Back out on the street, I peeked into the bag, and all the money was still there. I walked on air all the way back to Hotel Turismo. Unlike the bank teller, who probably lost his job for misplacing five thousand dollars of that ill-fated money, all I lost was a little pride for misplacing the sixteen thousand dollars that was left. Blame it on the *pisco sours* and the excitement of things to come.

The first place we stayed in the field was in a Catholic school for wayward girls, which was out for vacation at that time. It was located in the middle of Sombrero, the same small oil town a ways south of the Strait in the steppe of Tierra del Fuego where Erich and I had stayed on the earlier trip. Once we got there, I split the group up into three teams:

1. The zonal study team, which had to survey the entire oiled area to obtain a complete picture of the oil distribution, consisted of Mee-lace, Cristobal, and one Chilean scientist, Natalia. We would use one of the Argentine jeeps.

2. The oil team was to focus on the distribution and the chemistry of the oil in the most heavily oiled areas. This team, which was assisted by Alfredo and his trusty Volkswagen bus, consisted of Erich, Ana, and two Chilean scientists.

3. The tidal flat team had the tough assignment of studying the wide tidal flats with their fierce thirty-six-foot spring tides. This team consisted of Moon, Marcos, and one Chilean assistant, plus anybody else they could con into going out with them. They were to use one of the Argentine jeeps.

After the second day in the field, Tuesday, 10 February, I wrote the following in a letter to JM:

> Well, the oil is still here, that's for sure. Everywhere in great oodles and gobs. We've been spending a lot of time bouncing around over the hillsides and valleysides of the backwoods (so to speak; there are no trees) of Tierra del Fuego . . . I am now the world's leading authority on opening sheep fence gates. I am also pretty good at shoveling sand dunes out from under jeeps. Cristobal is the world's leading authority on getting jeeps stuck.

Next day, I stood on top of a cliff looking across the Strait of Magellan. I could just barely stand up, in fact, because the wind was eddying so strongly across the top of the cliff, possibly as much as fifty miles per hour. The water was a delightful green-blue color, and its surface was covered with whitecaps as far out as I could see. Well, that was it, nothing philosophical, it was just beautiful, that's all.

On Wednesday morning, our zonal team was driving cross-country in our jeep, as usual, trying to get to the beach. Cristobal did all the driving because I had not as yet obtained my international driver's license. I sat in the right front seat, and Natalia sat in the back, usually making critical remarks about our crazy maneuverings. We pulled up to the top of a high hill from which we could see the beach about half a mile away. Ahead of us was a very steep downward slope, with scattered vegetation. Cristobal looked over and asked, "What do you think?"

I gave a thumbs up, which meant go ahead down the hill, while Natalia said "Oh no, not again!" from the back seat.

We then lurched on down the slope, stopping abruptly about halfway. I looked out the window and noted that we were sitting up in the air, with all four wheels off the ground. Cristobal had driven the jeep on top of a huge shrub. The front wheels were six feet off the ground.

I looked around at Natalia, who was clucking an "I told you so" in Spanish.

All we had was a shovel and a small hand ax to cut down the shrub bush and dig out the jeep. Some of the limbs on that bush were eight inches in diameter. It took us about two hours. The wind velocity was its usual thirty-five to forty miles per hour, making it very uncomfortable under that bush in the blowing sand. Natalia refused to dig. It wasn't her idea to drive up on top of that tree.

I considered the next day, Thursday, 12 February, to be one of our best science days. We profiled a low-tide terrace 0.6 miles wide, and I learned a good deal about tidal flats in a macrotidal setting (which is a coast with a tidal range, the vertical distance between low and high tide, greater than twelve feet), something I had never studied in any detail before. Late in the afternoon, after finishing our low-tide surveys, we drove the jeep around the edge of the high marsh that bordered the large tidal flats called Banco Lomas that Moon was doing his thesis on. In places, these tidal flats are about five miles wide at low tide. When the tide turned, the team had to jog across the flats to keep ahead of the rising tide. Where the flats are widest is a big bend in the coastline, and the flats are muddy and soft at the head of the bend. Fortunately, only a small amount of oil was stranded on those particular mudflats and marshes. Farther around the coast by the First Narrows, the marsh at Punta Espora received a very heavy dose of oil, which, one and a half years after the spill, looked as if it had been deposited yesterday. The marsh at Punta Espora was one of the sites where Ana was concentrating her work.

We stopped the jeep and walked along the animal trails in the high marsh along the landward edge of Banco Lomas. We saw several *guanacos* (the national symbol of Chile), a huge deer-like animal with a long neck something like a giraffe, numerous geese, and one large puma-like cat. Walking through that ethereal marsh was like a visit to a world-class wildlife preserve, though the area was, in fact, part of a sheep farm. It felt as if we were intruding on the animals' secret sanctuary; therefore, we only stayed for a short time.

The sun shining at a low angle that late in the day flawlessly lit the plants and sky for the numerous photographs that we took looking toward the east. Unless you have been in the higher latitudes, it is impossible to appreciate the superb lighting that the low angle of the sun produces on a clear day. The pictures never do it justice.

When we changed shifts of the Chilean scientists after about ten days, the new crew brought the mail from Hotel Turismo, which included several birthday cards for Marcos. Nothing for Meelace, Cristobal, Ana, Moon, nor Erich. After reading the cards and making pronouncements about each one, such as "my dearest mother" and others I won't repeat here, Marcos taped them in a vertical row up the wall of the girls school. Then he set the lowest one on fire. As the flames leapt to the ceiling, we used the blankets from our bunks to save the school. Marcos, a handsome, wiry, blonde ex-quarterback, who looked a little like James Dean, was different, even for a geology major.

Speaking of geologists, one of the new Chilean scientists was a geologist who was dressed in a beautiful white suit and dark red tie when he showed up at the girls school. I asked what kind of geologist he was, and he replied "a field geologist!" So I sent him into the field the next day onto the mudflats with Moon and Marcos, a task that all the Chileans hated, usually asking to be transferred after one day. That was also true for this guy, whom I will just call "Whitesuit."

On 17 February, the zonal team, Meelace, Cristobal, and Natalia, crossed the Strait at the First Narrows and took up residence in Posesion, the oil camp run by ENAP on the Patagonia side of the Strait. The rest of the crew stayed behind in Sombrero, which was close to their study areas. We worked two tides a day, having some good days and some bad. By Sunday, 22 February, it was three weeks and three days since we left home, and except for Marcos, none of us had received any mail from the United States because of our remoteness and the difficulty of getting the mail to us from Punta Arenas. I sat down that afternoon and starting writing my sixth letter to JM, which follows, with several omissions:

Dear Jacqui—It's a Sunday afternoon in the old oil camp Posesion. It's a bright sunny day and the 30 mile-per-hour wind keeps rattling the tin roofs of the barracks and swooshing down the alleys between them with an uneven roar. We worked the early morning tide and will go out again in a while to make a long walk to a very inaccessible profile beneath a 200-ft cliff. Then we will be through with the eastern portion of the north side. We've worked both tides for the past few days and are right on schedule. It's been very windy and pretty cold most of the time, but there is lots of sunshine and there are many beautiful scenes on the beach. (I can't seem to write worth excrement these days. Sometimes I get inspired, but that fades before I can get out the pad. I seem to spend all my spare time— which isn't a hell of a lot when you're working two tides—reading.)

Anyway, tomorrow afternoon we will move to another oil camp called Punta Delgada, which is located further west, and, hopefully, have all the profiling wrapped up by the 1st. Then I can concentrate on some special projects and helping Moon on the tidal flats in the week or so that I have left.

Last night we just sat on top of that 200-ft high cliff and watched the sun set for a few minutes. It made some nice color patterns on the wide intertidal zone at low tide. As I watched, I huddled behind a dune to avoid the wind. In that protected place, it seemed so quiet. I could only hear the roaring waves—very still. I watched a single gull apparently having a great time soaring up and down over the edge of the cliff in the wind.

I guess every field expedition is different. So is this one. I don't know exactly what it is. For one thing, there is no talking. You can't talk with the Chileans and Cristobal is the silent type. Besides, I haven't been in the mood to talk. I like to talk sometimes. Talk. Talk. Talk. Talk.

Well, the moment of truth is rapidly approaching. Erich is going to Punta Arenas *tomorrow* to do some stuff and *pick up the mail!* Will I have a letter? Some letters? What will they say? I feel like I have been writing into a vacuum. Like I have a barrel over my head talking to myself. Are you there? Do you care? Does anybody really care? Hmmmm.

I guess I thought that my letters to you would take the place of my diary, so to speak. That I could write to you about all the wonderful and exciting and scintillating things that are happening to me. Truth is, not so many wonderful and exciting and scintillating things are happening to me. That is, that I can write about without getting horribly repetitive. For example, I love to watch the sun set and I've watched quite a few lately (big surprise!), but, obviously, you don't want to read one of my descriptions of a sunset in every letter. And everyday we go out and run profiles and measure some waves and look at some outcrops of till. We drive back and forth on the same dirt road and eat two meals a day in the same dining hall with the same 100 ENAP oil men and with the same 20–30 crack Chilean foot soldiers and with one lady who works in the kitchen and who looks like her face was run over by one of the Chevrolet pick-up trucks parked out front.

For some reason on this trip, I've been a little edgy, a little uptight. Worried a bit about the success of the project. Having all the extra Chileans here has been a real hassle. Also, too much is riding on too many green graduate students, to put it bluntly. And I can't spend as much time with them as I would like, because it takes so much time to do the zonal. Oh well, it's just something to worry about. Which is something I don't do very well, as you know. But, $90K is a lot of money (the amount of the NSF grant).

Two days later, we moved to the Tehuelche Hotel because the Punta Delgada oil camp was full. The hotel was in an isolated, two-story wooden building, owing its existence as a hotel to its location on the road between Punta Arenas and southern Argentina. Thinking back to my once in the museum in Punta Arenas, I knew that the hotel was named for an extinct Indian tribe, the Tehuelches, who used to run around naked on the hills of Tierra del Fuego. They didn't last long once the missionaries arrived.

The rooms in the hotel had no heat, so we spent our waking hours there sitting in the lobby listening to music. About 9:00 P.M., Erich called. He knew that the hotel was the backup if the ENAP camp was full. "Meelace, are you sitting down?" he asked.

At least that is what I thought he said. Although he was only a hundred miles away, the connection was so bad he sounded like he was on the other side of the world. "No, I'm standing in the lobby by the one and only phone in this place. Think I'm staying in the Hilton?"

"Well, Ana rolled her jeep, and she is now being held by the *policía* in Porvenir. She isn't hurt too bad, only a royal black eye. Two of the Chileans are in the hospital, though."

"Who?"

"Oh, Whitesuit and one of the new guys. They'll be out tomorrow."

"Great," I replied.

Then, after a long pause, I asked him to repeat everything, hoping I had heard wrong, which I hadn't.

"Okay, look, we only have a couple of more stations to do over here.

We'll be there tomorrow afternoon, as soon as we finish the stations. Anything else?"

"Well, Marcos is in the bed with a sore back and a bad case of diarrhea. Other than that, everything is about the same."

In the excitement of the moment, I forgot to ask him if he brought me any mail back from Hotel Turismo. When I got back to Sombrero, I found two letters from JM, and I read them over and over. She missed me, too, and she was still in love. What more could any man ask?

Next afternoon, I walked into the girls school quarters in Sombrero, and the first person I saw was Whitesuit. He was still wearing the white suit, which was streaked with mud from top to bottom as a result of his one day out on the tidal flats. He also had a terrific shiner and a broken arm. They told me that he had bounced around like a ping-pong ball in the back of the jeep as it rolled over in the gravel. He was a mess. Then Marcos came staggering out of the back room, not looking a whole lot better than Whitesuit. Eventually, Ana showed up, also sporting a black eye that looked like Whitesuit's.

It was time to regroup and make a new plan. Ana's jeep wouldn't be out of the garage for a while, not to mention the time it would take for all the broken arms to heal.

After a couple of days, things were more or less back to normal. Right after dinner every night, we had a science meeting at which we reviewed the results of the day's work and made plans for the coming day. At the end of one of the meetings, I asked for any final comments.

Marcos spoke up. "I know that I don't have to tell you this, Meelace, but we are running way behind schedule on the tidal flats project. Eduardo is the only Chilean who has stuck with us. And it takes almost a full tidal cycle to walk out to the low-tide line and back. It's getting tough! However, I have a plan."

"Oh yeah, Marcos, what is it?" I asked, noticing some skeptical smiles around the room.

"We need a helicopter to get all the way out there to those sandflats!"

As mentioned previously, the average distance to the low-tide line was four to five miles.

"No kidding. And just where would this helicopter come from? Have you seen any around these ENAP camps? I doubt if there is a single helicopter in all of southern Chile."

I was losing my patience. It was late, and we had an early morning tide.

"Sure there is. There's two right over there in Punta Arenas!"

"You're right. I forgot. A minor detail, however. They belong to the Chilean navy. I was told that they are the only ones they have in the whole country."

"Let's use them," he said, warming up to the idea.

"Just how do you propose to pull that off?" was all I could think of to say.

"Well, I'll catch the plane back over to Punta Arenas tomorrow, and then I'll go see the admiral, blink my baby blue eyes at him, and ask for the helicopters. It'll be easy."

I thought about it for a while, amid the laughter of the others. "Okay, go ahead, what have we got to lose? Otherwise, we'll never see the sand-flats except from the air in the fixed wing."

Everyone, except Marcos, was dumbfounded. They knew that I had an uncommon tolerance for Marcos's unorthodox ways, but this was going too far.

Next day, as he suggested, Marcos flew to Punta Arenas. The following morning, not too long after sunrise, we heard fltt, fltt, fltt, fltt, fltt overhead and ran outside to see two bright orange helicopters circle the school and go in for a landing at the airport.

We used the two helicopters, which were piloted by American-trained Chilean pilots, for three or four days, completing Moon's work on the sandflats with alacrity and overflying the entire study area. A minor problem developed when, while making a seat adjustment to better shoot photographs, I accidentally shut off the engine of one of the helicopters over the middle of the widest part of the Strait. We only dropped a few feet before the pilot righted the situation. He later said, "I was never so scared in my entire life."

During the field study in Chile, we established ourselves as a team with *bona fide* oil-spill experience, although admittedly, we had not yet been to our first "live" oil spill. In summary, we learned a great deal about the distribution and weathering of oil in a number of coastal habitats, especially gravel beaches and salt marshes. These days, when lecturing about this particular spill, I refer to what I call "the first rule of oil-spill behavior": *salt marshes are susceptible to long-term impacts from oil spills,* which we first learned at the *Metula* spill. We always give salt marshes the highest possible rank in sensitivity during our mapping projects. Most modern oil-spill contingency plans focus a significant part of their protection strategies on salt marshes. I do not wish to imply that we were the first to think of this concept. Other people were working on oiled marshes well before us. However, the Punta Espora marshes by the

First Narrows of the Strait of Magellan still today represent a classic case of salt marsh oiling. NOAA scientists visited the site twenty years later, and the impacts of the oil could still clearly be seen.

Through a series of long-distance phone calls, we made arrangements for JM to visit the field site for a few days. She arrived for the end of the helicopter survey. When the project was over, she and I went to Bogota, Colombia, for a couple of days, where we contracted diarrhea so bad we almost had to rent separate rooms. From there, we went directly to St. Croix in the West Indies, where another group of my students were conducting a research project. After the reunion in Chile and the stop by St. Croix, the die had been cast for JM and me. This was the real thing; I was definitely no longer "in between dances."

The Black Tide of La Coruña

Having thus lost his Understanding, he unluckily stumbled upon the oddest fancy that ever enter'd into a Madman's brain; for now he thought it convenient and necessary, as well as for the increase of his own Honor, as the Service of the Publick, to turn Knight-Errant, and roam through the whole World arm'd Cap-a-pee, and mounted on his Steed, in quest of Adventures.

Miguel de Cervantes, *Don Quixote*

Sunday, 16 May 1976—
Between Bangor, Maine, and Boston

We took off from Bangor on this bright, clear Sunday morning over the partly forested hills of central Maine. We were not quite far enough north to escape the purple haze from the cities to the south that hung at two thousand feet.

We had been working the shoreline of Maine for ten days when I noticed a short article in the newspaper about an oil spill in the harbor in La Coruña, Spain, which is located on the Atlantic coast at the northwest corner of that country. A quick call to Chris Ruby in Columbia confirmed our suspicions. It was a big one, and we had to go. Therefore, Erich Gundlach and I were riding along in a jet on our way to our first "live" oil spill, leaving JM and the rest of the field team to drive our vehicles back to South Carolina.

Out of the window to the left, I could see the islands we had touched here and there on our once-over-lightly survey up the coastline of Maine. I thought, "Where will the oil go when the tanker hits the rocks at Eastport? Onto the rocks, onto the tidal flats, or into the marshes?"

Occasionally, I would think ahead to the spill. Chris told us he had seen pictures of waves of oil breaking on the Spanish beaches on the evening news. We had great expectations. What would the northwest shoreline of Spain look like? And Madrid?

We flew over the rest of the coast down to Boston and then on to New York City in the fog where we sat around for hours waiting for airplane tickets and passports. Finally, we were off on an all-night ride in a crowded Iberian Airlines B-747, with Rooster Cogburn and cigarette smoke for company.

I woke up and looked down on top of bare mountains and wide valleys covered with a quilt work of cultivated fields surrounding the clusters of houses in the small villages. The landscape looked old and stagnated, with underfit streams in older flood plains. The sun was high. My watch read 12:30 A.M. It had been a long and painful trip. Thank God for Rooster Cogburn.

On the approach into the airport at La Coruña, we saw the half-sunken tanker, which was named the *Urquiola,* and the oil scattered in pockets all along the beach, streaking in slicks away from the anchored boats and, finally, coating a solid black the surfaces of the intertidal areas. We concluded we had made a wise decision to gamble on this trip. I thought, "This spill will give us a good idea of what is in store for the shoreline of Maine and southern Alaska in the not-too-distant future."

We rented a car and drove down to the port, which is guarded by the *mas antiguo faro en todo mundo,* the oldest lighthouse in the whole world. A black smudge of oil coated the high-tide line of the entire port.

On the following Tuesday afternoon, I had my back to the seawall at the landward end of one of our survey lines, squishing in the oil with my waders as I tried to sketch in the exact location of the oil ponds, the beach cusps, and the point where the *marea negra* (black tide) had broken over the seawall. The tidal range in that area is eight to ten feet, and we were, as usual, working the low tide. I heard the voices of two young women. As I turned back to the seawall, I looked up at one of them, who had walked up to the edge of the wall.

"*Buenas tardes!*" We exchanged greetings. She rattled off some words I didn't understand and started waving her arms in the air. I stared back at her dumbly.

"She wants to know when she can swim on the beach again," Erich called up from down on the beach.

"*Oh, muchos días* (many days)," I finally answered.

"*Ah, muchos días.*" She nodded.

"*Si, muy malo,*" I replied. They both turned to walk away and then turned back to ask if I was from Holland.

"*No. Estados Unidos.*"

Later, at the same station, I saw two old women walk by with handkerchiefs over their noses. I was inspecting our whisky bottle full of oil at the time. We had collected it out of a breaking wave in the surf. The sample looked like it was ninety percent petroleum. The substance in the bottle was what oil-spill responders call *mousse* because of its resemblance to chocolate mousse, a dessert. *Mousse* is a water-in-oil emulsion, which may contain up to eighty percent water. Therefore, after this emulsion process takes place, which is aided by wave action, the volume of the oily substance that needs to be dealt with increases by as much as a factor of five.

We knew already that you were supposed to collect oil in precleaned and sealed chemical sample bottles, not discarded whisky bottles, but we had forgotten to obtain bottles during the mad dash to New York-Madrid-La Coruña. Give us a break—this was our first "live" oil spill.

Then we went to a sand beach where we met the chemical oceanographer from the local Oceanographic Institute, who was digging a trench in the beach with a small stick, looking for subsurface oil and shooting his one and only roll of film. We may not have had sample bottles, but we did have about twenty rolls of film and a couple of shovels, with which we found subsurface oil at most of the stations we surveyed, except those composed of solid rock. And shoot pictures we did, with the photos turning out to be the best I have ever taken at an oil spill. To this day, slides taken at that spill are featured in a number of the lectures in our oil-spill training courses.

The chemist had looked through the window of his laboratory as the tanker exploded and burned, and the oil seriously impaired the research they were carrying out in the saltwater ponds attached to the facility. He had trouble shifting mental gears to carry out a research project on the catastrophe going on all around him, instead of the other projects already in progress. Besides, his government wasn't interested in paying for oil-spill research, he said.

I made some derogatory comments about that lab's approach to science in my original notes, but I think I won't include them here. After all, you now know that we collected an oil sample in a discarded whisky bottle.

That night, I was standing by the window of our room on the eighth floor of Hotel Riazor. In daylight, you could look down from that window and see the oil swashing on the beach and the detergized streaks of the "chocolate mousse" in the surf. But it was dark, and I was listening to the familiar surf sounds, mixed with the auto sounds of the road. I thought, "The oil is a phantom going here and there at its own bidding to strike silently in the night while we sleep, leaving a trace of black poison wherever it goes. What new species will we see washed up in the swash lines of the beach tomorrow? Because there are no blood stains on the door post to bar the entrance of this angel of death, this *marea negra* that will call again on the mudflats, marshes, and rocky tidal pools of Ria de Betanzos, Ria de Ares, and Ria de el Ferrol del Caudillo tonight."

One reason the photos taken on this particular trip were so valuable was because of the variety of habitats that were oiled in the estuaries, which are called *rias,* in the spill site. We had seen a major kill of burrowing, edible clams (cockles) on one of the sandflats, and the photographic documentation of that kill has been cited often in the scientific literature.

Next morning, Erich dropped me off at the airport in Santiago so I could go home to make arrangements for a more complete study of the spill. I was a little late getting in line, so I walked alone across the concrete toward the waiting B-727. The sun was just rising behind me, and a faint yellow-pink band streaked across the sky behind and above the plane. Fog hung scattered over the trees in the background.

I felt a little overwhelmed to be so involved in such a historic, newsworthy happening. After all, the *comandante* of the Spanish coast guard was going to fly my crew in his personal helicopter that very afternoon. Ana and Cristobal were on their way, and we seemed to be the first science team on the spot. In my pocket was a letter of intent of cooperation from the *jefe* of environmental pollution in Spain, who was gung ho for the study. Everything seemed to be falling into place, like a dream come true.

Four of us left Columbia on 1 June and headed back to Spain. Cristobal met us in Santiago, and as we rode along toward La Coruña, we made plans to become the world's greatest oil-spill response team. Here's the picture. Forty-one year old Knight-Errant Don Billy Quixote sets out to save the world with his faithful servant Pancho Hardnose (Erich) in a Boeing 727. Anyway, I felt like Don Quixote as I splashed through a heavily oiled marsh that we previously thought had not been affected at all by the oil.

A little later, we walked away from the oiled marsh up a flowered hillside around the edges of several planted gardens in an ancient Spanish field. It felt as if centuries were passing under my boots, which were covered with the oil from the twentieth century industrial revolution that would soon bury all the years.

On the way back to the hotel, I watched the old men and women plowing the fields with milk cows and saw the lumberyards, cement plants, and granite quarries. And the tight pants of the *lolas* and the fifteenth-century churches. A century of time collapsed as the old women dressed in black, the black skirts of Galicia, stood around waiting to die.

Next morning, we visited the circus at the gravel beach at Mera, the hardest hit beach, where clean-up procedures were still under way. I saw untold numbers of older men dressed in suits and ties, presumably bureaucrats and politicians, come by and look, wave their arms, talk a lot, and then leave. But there were only six fishermen-out-of-work down in the oil trying to clean it up.

On our first visit to Mera, ten days earlier, a crew of three old boys from Louisiana representing Oil Mop, Inc., was there dragging something called a "rope mop" through the oil. It consisted of what appeared to be a bunch of janitor's dustmops strung along a rope connected to a pulley, which was attached to a buoy offshore from the beach a ways. They stood at the waterline as they manipulated the rope. When the mop came back up out of the oil, they squeezed the oil out of it into a container and then passed it back toward the pulley. This was a rather smallish operation, to say the least.

They greeted us in a friendly fashion, and being from South Carolina ourselves, you might say we spoke the same language. After we introduced ourselves and told them why we were there, one of them said, "Y'all study it, and we'll clean it!"

I looked at the large pool of mousse being held against that pocket beach by an onshore breeze and estimated that maybe about five thousand barrels of it were on the surface of the water. I asked, "How much oil have you collected so far?"

"Oh, about a barrel."

"Only about 4,999 barrels to go," I thought. We wished them the best of luck and went on about our survey.

In the afternoon, we stood on a hilltop looking down with our long lenses to where several men were raking up the oil with flat wooden rakes. One of the workmen, covered as he was with oil from head to toe, waved to Ana to come on down and join in the fun.

We outlined the principles we learned at that spill in a paper, with Erich as the senior author, published in the journal *Environmental Geology* in 1978. In that paper, we estimated that twenty-five thousand to thirty thousand tons of oil washed onto the shorelines at the spill site and provided details on how the oil impacted the different habitats. When I talk about this spill today, I refer to two more rules of oil-spill behavior that we learned there.

The second rule of oil-spill behavior is *on exposed, high-energy rocky coasts, the cleanup of spilled oil by natural processes is rapid (hours to days)*. We observed the waves reflect off several rocky cliffs on the outer coast, creating a current that held the mousse offshore. Therefore, the cliffs were not materially affected by the oil.

The third rule of oil-spill behavior is *the depth of oil penetration and burial on beaches increases with increasing sediment grain size*. At the *Urquiola* spill site, a wide range of beach types were present, including gravel beaches, like those so common in New England and Alaska; coarse-grained sand beaches, like those on the outer shore of Cape Cod; and hard-packed, fine-grained sand beaches, like the ones on the barrier island shorelines of Texas. Oil penetrated to depths of two feet in the gravel beach at Mera, but only to less than an inch in fine-grained sand beaches on the outer coast. Rapid changes in the beach cycle promoted deep burial of oil to depths up to three feet on the coarse-grained sand beaches. These findings have been of great value to us in the design of cleanup strategies for the oil spills we have responded to since that time.

My last day in the field was spent in Portugal, looking over the year-old oil-spill site and wreck of the tanker *Jacob Maersk*. Oil was still on some of the rocks, but most of it had been washed away by the waves because of the site's exposure to the open ocean. The Canadian scientist who wrote a paper on that spill called it "a very fortunate spill."

Next day, I was back on a series of jet planes, Oporto-Lisbon-New York-Atlanta-Columbia. When I got home, I learned that the Spanish government had decided the *marea negra* could use a little science after all and had dispatched over twenty scientists to La Coruña for a concentrated study.

Shortly thereafter, our crew was asked for political reasons to cut short our work and leave the country. Yankee go home!

On Spain

In 1492, Columbus sailed the ocean blue—
In 1521, there went old Magellan—
They sailed the seas for Spain,
In search of wealth and fame.

Then came the Conquistadores—
I read it in some stories—
About how they took the gold,
These brave bold men of old.

In 1976, Billy Whiskers tried to mix
Some science with black-tide ecology
And even some geology!

But the Spaniards calmed him down
When they asked him to leave town.

Billy Whiskers

1976

"Oh boy—I sure do like the sunshine."

Drunken Eskimo in Gold Nugget Bar,
Kotzebue, Alaska

Early June 1976—
Lower Cook Inlet, Alaska

"Hey Mo, Red says he wants to go now. The tide's coming in." That
was Jeff Brown, another one of my graduate students, who, along with
JM, was assisting with a survey of Lower Cook Inlet.

From where we were standing, which was on a wide intertidal sandflat on the east side, you could see clear across the inlet to two of the
volcanoes on the west side, the massive snow-draped Mt. Iliamna and
Mt. Augustine, the latter still smoking from its latest eruption.

"What's the hurry?" I finally responded, "looks like the tide's still a
long way out to me. Did you see these rock outcrops under the gravel?"

Not that I wasn't concerned about the tides. The tidal range at the
spot we were working was about twenty-eight feet, but I had memorized
the tide table and knew we had a few more minutes. Noting that I wasn't
planning to leave anytime soon, Jeff ran back to the plane to consult with
Red, the pilot.

Somewhere along the line, I had managed to land a small contract
with the Alaska Department of Fish and Game to conduct this survey.
Pat Wennekins, a member of that department, was concerned about the
possibility of oil spills in Lower Cook Inlet, which was connected by a
wide channel with Upper Cook Inlet, an area of active offshore oil exploration at the time. He contacted me when he learned about our work on
the *Metula* spill in Chile and the similarity of the Strait of Magellan to
Cook Inlet. Or was it the other way around?

Believe it or not, I proposed to do a zonal study of the area, which meant surveying numerous sites all the way around the inlet, before coming up with useful information for a contingency plan for oil spills, should they occur there. My theory was that you had to know what would be impacted before you could protect it. The focus of our work was the shoreline habitats. Pat agreed with my plan and said "have at it." Our fee was a whopping fourteen thousand dollars, which was fine. We weren't doing it for the money then; we just wanted enough cash to cover the costs.

Anyway, the three of us showed up in Homer, Alaska, which is located at the entrance to Katchemak Bay, an arm of Lower Cook Inlet located about halfway down the east side. We rented a tail-dragging Cessna 180 piloted by a young red-haired pilot from California. This particular day was our first day out with Red, and things were not going smoothly because he steadfastly refused to land where I told him to. It was only after much cajoling that I finally convinced him to land on a long, low intertidal sandbar on the sandflat we had just finished surveying. He was very nervous about it. The station was actually a wide, wave-cut rock platform with a thin sand veneer, a technical detail that I'm sure Red never noticed.

I was down on my knees examining a rock outcrop near the high-tide line, when Jeff came running back, this time accompanied by JM, who said, "Red says he is going to take off right now if we don't come on."

The plane was about thirty or forty yards out on the flat, and I walked slowly, grumpily back to it, even taking a couple of pictures along the way. One of those photos, which shows the plane framed against the background of the volcanoes in the distance, now hangs on the wall of our business establishment in South Carolina. The water was still fifteen or twenty yards from the plane. But by the time I got there, Red was out of the plane pointing at the tires, which had sunk a few inches into the sand because the water table was rising as the tide came in. I started to suggest a solution to his dilemma, but before I could say anything, he jumped back in the plane and started cranking the engine. When I climbed in the front seat beside him, he was furious.

"You SOB, you are going to cause me to lose this plane and my job. I knew I shouldn't have listened to you. I knew better than to land here."

Meanwhile, I had the window open, looking down at the partially sunken right wheel. Before I could say anything, he revved up the engine with a loud roar, and the wheels moved not an inch.

"Stop!" I cried.

But he ignored me, revving it up to full throttle a second time. I was afraid he was going to blow the engine. Finally, I got his attention, and he cut back on the throttle.

"Red, I've been in this situation lots of times before. Let the three of us get out, and we'll push on the struts and rock the plane a little bit. When we do that, you give it some throttle, it will pull the wheels out of the sand, and you can roll down the bar a ways. Then we'll get back in the plane, and you can take off. It will take a while for the wheels to sink in again."

After another round of swearing and cursing at me and another attempt to blow up the engine, he reluctantly agreed with my plan. We got out and pushed on the struts while he engaged the throttle again, and the wheels rolled easily out of the divots they had made in the bar. He rolled on down the bar, but he didn't stop like I had suggested. He took off.

"Huh oh," the other two said almost in unison, "He's not going to come back for us."

We were facing a thirty-five-mile walk back to Homer. At least that was how far it was as the crow flies. We may have been able to hitch a ride if we could have climbed the high cliff behind the beach to a road somewhere.

While Red circled round and round above us and contemplated what to do, we walked farther toward the high-tide line to find another landing spot. By then, the bar the plane had been on was almost under water. We made a huge arrow in the sand, implying that he should land there, which he finally did, still fuming and cursing.

When I got back in the plane, he said, "You SOB, I will never fly with you again as long as I live."

"Well son, what makes you think I would want you to fly me again? You just lost this contract."

Back in Homer, I asked Red's boss if he had any pilots who didn't mind landing on beaches, and the boss said his company usually discouraged beach landings. He said I wouldn't find any other flying services in Homer that would take on that kind of responsibility.

"Well, where can I find somebody?"

"Maybe in Kenai."

We walked into a flying service in Kenai, which is about fifty miles up the east side of the inlet, where we were met by one of the pilots.

"You land on beaches?" I asked.

He grinned a little and answered, "Now and then."

He had a slight build and wore a brown cowboy hat and cowboy boots at the bottom of his bowed legs, looking as if he had just stepped off the set of a western movie. I noticed that he had a major scar that cut across his face over the bridge of his nose.

"Well, I guess you're our man if you want the job. We have to land on about fifty beaches all around the inlet. I'm Miles Hayes, from the University of South Carolina; this is Jeff and Jacqui, also with the university."

"Sounds interesting. My name is Pierce Bassett, but most people just call me 'Cowboy.'"

Next morning, we took off from Kenai, and he spent quite a while circling to gain altitude. "Uh, we want to go over to the other side," I said, thinking that maybe he had not listened to my initial instructions.

"Yeah, just be patient, we'll go on over in a minute. You have to be up at five thousand feet in order to glide to the other side from the middle. Guess you noticed that we have only one engine."

"Good plan," I responded. I was already liking this guy.

At the first station on the west side, another wide sandflat much like the one we landed on the day before, he went straight in and landed without a moment's hesitation. "Don't you check these places out before you land?" I asked.

"Don't have to check this one; I land here almost every day. We service a group of salmon fishermen who stay in that cabin over there."

After finishing that station and one more very similar to it, we came to the front of the Red River Delta, which has steep beaches composed of coarse-grained, sometimes very soft sand. "Land there," I pointed to an arc of the beach on the delta front as we passed over it. He circled back.

"Never landed here before. I'll have to check it out. See any cables?"

"Can't say that I did. Why?"

He pointed to the scar across his nose. A couple of years earlier, he had not seen one of the cables to which the salmon fishermen attach their nets. The cable was up a few feet above the beach when Cowboy came in for a landing. When he hit the cable, the plane flipped over on its top, giving Cowboy his permanent scar.

Then he went straight down and along the beach, just touching the right wheel to the beach as we flew south. He would vary the weight on the wheel as we flew along, almost bouncing the plane. Next, he circled back low over the spot where he had touched down, checking how much the wheel had sunk into the sand.

"Looks okay, but a little steep," he said, and then he circled around and repeated the one-wheel performance for a while before he gently lowered the other wheel down on the steep beach, coming to a stop exactly at the spot where I had originally pointed.

After that performance, I knew we were flying with somebody special. With all due respect to Pat James, Jim Foote, John the Pilot, "Mudhole" Smith's son, Gayle Rainey, Doug McCart, and all the others, Cowboy Bassett was by far the best bush pilot we ever flew with. He was an artist with an airplane. He could have landed it on a diving board if it were long enough.

The rest of the Lower Cook Inlet survey went smoothly. The result of that survey was a mapping concept that I first called the "vulnerability index," in which shoreline habitats were ranked on a scale of one to ten based on their sensitivity to oil-spill impacts. This original ranking was based primarily on the observations we made at the *Metula* and *Urquiola* spills. With the help of biologists Geoff Scott and Chuck Getter at RPI, biological components were later added to the concept, which is now known as "Environmental Sensitivity Index (ESI) mapping." This kind of map allows oil-spill responders to make decisions regarding the utilization of equipment at a spill because greatest attention can be directed to those shorelines that are most sensitive (that is, those habitats with the highest ranking on the maps) or to areas of greatest concern for the biota. Wetlands, such as salt marshes, are usually the highest-ranked shorelines. By early January 1997, twenty years later, ESI mapping had been conducted by us mostly for NOAA and for some state agencies of most of the U.S. coastline. These maps are used as part of the planning process at all sizeable oil spills that occur in the United States. The ESI concept has also been used, by us and by others to map shorelines in other countries as well, such as Canada, Kuwait, Germany, and several countries in Southeast Asia. Much of this work has been government sponsored, but as I was writing this, JM gave me a 1996 publication by the International Petroleum Industry Environmental Conservation Association (IPIECA) entitled *Sensitivity Mapping for Oil Spill Response*,[27] which recommends an approach that essentially adopts our ESI technology.

On 20 July 1976, Chris Ruby and I met Cowboy in Anchorage for a flying examination of the entire Alaskan shoreline between Anchorage and Prudhoe Bay, even to the Canadian border. The purpose of that trip was to acquaint ourselves with the coastline that we would later map using the ESI concept. We have now mapped all of the Alaska coast, except for a segment along the Chukchi Sea, which was mapped by a competitor, and a part of the far southeastern region. In fact, that was what we were doing when I endured the plane crash described in the first chapter of this book.

The first leg of the grand tour of the Alaskan coast in 1976 took us to a large, military landing strip at Cold Bay, which is located near the very end of the Alaskan Peninsula. I wrote a letter to JM from there, which started "Cold Bay is where I am and cold and bleak is what it is."

During a visit to the Weathered Inn, a bar in Cold Bay, I asked the only person in sight, a middle-aged man named John, who I guess was a regular at that establishment, what went on at the base in Cold Bay.

"Doing classified work is all I can tell you," he replied in a raspy, broken voice.

It seemed that almost everybody in Cold Bay was in the last throes of losing their voices. There must have been a bad shipment of whisky somewhere along the line.

That same night, John started a minor ruckus in the bar by calling the local cop and telling him the place was being robbed. The angry, rolly-polly policeman threatened to throw John into jail. "The least he could have done was buy us a drink," was John's only response to the threat.

John, who had a beer belly about three yards around, said he used to be a pilot. "I can fly better than any of these farklers," he asserted, "I just don't want to die in an airplane." Made sense to me.

Next day, we took a seven-hour round-trip down to Dutch Harbor along the southern shore and back to Cold Bay along the northern shore. Unimak Island, the first large island in the Aleutian chain, was a geological wonderland composed of four large, steep-sided volcanoes, which looked like four inverted ice-cream cones, smoking at the top. Fresh lava flows came right down to the water, to be carved by the waves into sea stacks, sea arches, and sea caves. Would you believe that I took a lot of pictures?

After spending another exciting evening at the Weathered Inn, we headed around the shoreline of Bristol Bay, past the Bering Sea coast of the Yukon Delta, and eventually ended up at the town of Kotzebue, located just above the Arctic Circle. Along the way, as an added treat to all the observations and photography, Cowboy taught Chris and me to fly the plane. I forgot to mention that one of Cowboy's skills was aerobatics, which he demonstrated to us from time to time with some spectacular rolls and dives. One day after we had completed a section and had to double back for fuel, Cowboy told me to take the controls of the plane.

He said, "Now close your eyes, keep them closed tight, and see how well you can fly blind."

We were flying at a pretty good clip, and I thought I was doing okay, although I did hear some strange noises coming from Chris in the back.

"Now open your eyes."

I did and discovered we were flying upside down.

The shoreline of Kotzebue Sound was to be the subject of our next Alaska field survey the following summer, with Chris Ruby running that project. Therefore, we spent some extra time in Kotzebue.

After one long day of flying out of Kotzebue, we dropped in on the

Gold Nugget Bar. I was talking to Mary, the bartender, when another Eskimo lady sat down beside me and asked me to buy her a drink, which I promptly did. The place was full of drunken Eskimos, and one cowboy and two oil-spill diggers, who had not so recently returned from the steppe of Tierra del Fuego, were also pretty drunk.

"And why are you laughing about the native peoples?" she asked after she had finished the beer.

I tried to explain that I was laughing at my own native peoples, whoever they were, and not hers. In the past year or so, I had seen too much of how the people who lived with the land like they were a part of it had been driven to an early grave. These observations would now include the Eskimo peoples of the Arctic Circle. Even the land itself was being consumed at a catastrophic rate by the tidal wave called "progress."

She didn't know what I was talking about. Did I?

After she left, I looked out the window to where the fog obscured the salmon boats that rose faintly on the horizon as the swell rolled in to break on the debris-lined wall of Kotzebue, which sits precariously on a long spit that reaches out in the fog to touch the Chukchi Sea.

Chris was talking with one of the old Eskimo men. Chris told me, "he says the ice washes over the spit in the spring, and the people run to hide on the land in their igloos."

"Guess we'll find out for ourselves next spring," I responded.

Next stop was dreadful Point Barrow, which made Kotzebue look like the French Riviera—more native peoples living in poverty. And then on to the oil camp at Prudhoe Bay, where you could get anything you wanted, including a rainbow trout dinner.

After reaching the Canadian border, we turned south toward Fairbanks, with Cowboy letting me "fly" the plane across the unbelievable Brooks Range, another visual treat.

In November 1976, we started the waterfront project in Kuwait that I alluded to earlier (in the chapter "A Primer on Oil Spills"). Working with the architects of Sasaki Associates of Boston and Kuwait, and the engineers of the municipality of Kuwait, I was introduced to a new type of clientele: people who wanted answers, real answers, to their problems and who did what you recommended they do. This was exciting for me, a sneak preview of what I would encounter later in the real business world.

I don't believe I ever had a more productive year, scientifically, than 1976. I hadn't been in that mode since 1969. By the way, you may have noticed that this chapter contains no poems by Billy Whiskers. Know why? Because Billy Whiskers was now as extinct as the Tehuelches.

Cordova Revisited

Sunday, 16 January 1977—
On an airplane from Columbia to Juneau, Alaska

I was heading for Alaska on a marketing trip. It was frigid cold on the east coast and in the Midwest. The pilot said the chill factor on the ground was −41°F in St. Louis when we passed over.

After takeoff in the fog from Seattle, I looked down at the mountain peaks rising above the fog blanket and thought "I'm glad I was able to live on the earth in a time when I can fly through the air and see it all. And that's exactly what I want to do, at least most of it, before it is all logged off or otherwise completely changed." Just at that time, the bald, clear-cut blotches on Vancouver Island were coming into view.

A day and a half of marketing in Juneau more or less consolidated the following summer's work on the Alaska coast, but I was depressed by the meetings, mostly because I gave a lecture on oil spills that everybody slept through.

Once out of Juneau, however, we started flying along the old familiar south-central coast on our way up to Anchorage, which was a real lift to my spirits. I watched most of the area we had studied in the summer of 1975 pass by, including the chalky white Mt. St. Elias and Malaspina Glacier, covered with their new layer of snow. No lights were on in Cape Yakataga.

In my hotel room in Anchorage, which was in the throes of its warmest winter on record, I heard a politician on TV say he was sure that the port of Valdez could be made totally accident free. (This statement is taken directly from my diary notes of that date.) The news commentator noted that an oil tanker was sinking somewhere in the middle of the Pacific. He stated further that probably not very many people were interested in that fact, "especially the people freezing on the east coast."

After another day of meetings in Anchorage, I walked out of the hotel at 6:00 A.M. on 20 January, Jimmy Carter's inauguration day, and boarded the first yellow cab in line. The driver, who had long hair and a beard, had been writing on a piece of paper just before I got in. I glanced over the seat back and saw that he had written "NO DOPE/NO HOPE/NO RIDE/I DIED." I was going to Fairbanks for more meetings and would return south on Friday night.

In the airports in Anchorage and Fairbanks, the aura—the sense, taste, and smell—of the pipeline was all about me. I could see the urgency of the northward migration in the faces, and hear it in the conversations, of the passengers still heading out in the Great Alaska Oil Rush.

Friday night I caught Alaska Airlines flight number 96, the "Pipeline Express," between Fairbanks and Houston. I was sitting next to a six-foot-plus man with a highly wrinkled brow above large bloodshot eyes, which were partly hidden behind his oval-shaped glasses. He had a huge beer belly hanging over his Levi's and wore cowboy boots and a yellow-and-white baseball cap with "Trans-Alaska Pipeline Project" printed on the front. He looked extremely tired, but he seemed to want to talk.

He said he worked twelve hours on and twelve hours off for three weeks straight and then went home to Texas for two weeks of R&R. He made twenty-one hundred dollars a week.

"In the past two years, I've made one and one-half to two hundred thousand dollars, and I don't know where it all went," he complained. "I'd like to take a couple of weeks off and do some camping and fishing, but I don't ever have time to do that."

"Know what you mean," I commiserated.

"I've got my own banking account up here now, and I put half of the money in it. I don't tell my wife how much." Up until he made this move, his wife had been giving him an allowance out of his two hundred thousand dollars.

"You need a new bookkeeper," said I, thinking about how helpful my accountant, Jim Davis, had been.

In the fall of 1974, after I had been on my own for a few months, I was not exactly spending my money wisely, to say the least. My resources were quite thin because I had left Massachusetts with a flat bank account and what I could pack in a U-Haul trailer. One day I realized that I couldn't even buy groceries for the coming week, not to mention eat out in the different restaurants around town as I had been doing. I hadn't been that broke since I was a student at Berea College. Faced with that harsh reality, I hocked my *cameras.* For peanuts. What?

I consulted with my friend Sally, who was divorced with two children and doing fine on her own. She said, "you need a bookkeeper" and recommended Jim Davis, who put me on an allowance, and before long, I was taking Sally out to eat again.

Back on the "Pipeline Express," which by then was flying in the dark over the spot where I would some day have to swim to a rock and wait for Jamie to find me, or somewhere along there, I asked, "Think we're going to run out of oil in thirty-five years, as the pundits say?"

"Hell no, there's no oil shortage! One well was drilled for nothing."

Not knowing exactly what that meant, I changed the subject to something I knew a little more about—namely, the environment. He then told me that one of his co-workers collected blue fox furs,[28] which were apparently valuable.

"He puts grease on an iron stake. The fox licks it, and its tongue sticks to the stake because its so cold. Then he knocks the fox on the head. Saves him from having to run around in the snow and ice to find them. He has collected twenty furs using this technique."

"Umh huh," I coughed.

"Hell, I never had no education. I quit school and went to work."

"Sounds like you're doing pretty good to me."

"Yow, you have to do the best you can with what you've got."

Truer words were never spoken. And the two brothers under the skin talked on into the night over the whir of the engines carrying them south and to home.

At the beginning of April, I was back in Cordova, Alaska, serving as a panelist for a meeting, the general theme of which was the potential for the new pipeline's environmental impacts on the town and surrounding area, including the tanker traffic that would soon be heading by there on its way to the Valdez oil terminal. I was probably recommended for the panel by Pat Wennekins, who thought the Lower Cook Inlet report with its "vulnerability index" was a good idea. Pat was in the audience, as was Nancy Maynard, a Ph.D. biologist with NOAA, who was following our work and would eventually represent NOAA at many spills. The meeting was chaired by Rudy Engleman, also of NOAA.

Though there is no record of it in my notes, I presume I talked about the coastline around Cordova, which, as you know, we had studied in some detail, and added a little something about our oil-spill experience.

Clayton McAuliffe was there representing Exxon, and he gave his usual industry apology speech, showing pictures of fish living "happily" around the offshore oil platforms in the Gulf of Mexico. I believe that is where I first heard him make a statement I will never forget: *"It has never been proven that oil spills do any harm to the marine environment."*

That's a matter of interpretation, as I have already mentioned, but when I heard it then sitting up in front of a large audience of local townspeople, including some of my fishermen and pilot friends, and a number of the native peoples, I cringed, thinking, "Guess Clayton never slopped through the oil in the Punta Espora marshes or walked on the dead clams in the *rias* of Spain. If he had, he could never make a statement like that, or could he?"

At last came the moment of truth for the panelists, the floor was thrown open for the townspeople to ask questions of the panel, and a dozen hands shot up. Rudy recognized a native lady sitting in the third row. Looking out across the glare of the stage lights in front of the podium, I could barely make out her round, tan face in the background. She stood up, looked directly at me, and said, "I want to know the personal opinion of the people on the panel about what will happen if we don't stop this." She paused, with a look of puzzlement crossing her face, "I know you depend on this thing for your livelihood, but what is your personal opinion? What if we do have a catastrophe, what then?"

Rudy waffled, "Anything I say has to represent NOAA. There are reporters in the audience. I'll talk with you later."

"That's a cop out!" One of the fishermen farther in the back.

"Oh come on Rudy, that's BS!" yelled one of the native men.

I sat silently looking at Pat and then over to Nancy and then back at Rudy. I was almost ready to give my very own personal opinion, just as the native lady had requested. I was going to say that Clayton was a walking Exxon commercial. That the Gulf coast wasn't exactly as pristine as the pictures of the "happy" fish by the offshore platforms seemed to imply. I was thinking about the tar balls on all the beaches and the ripped up marshes behind the barrier islands, much of which had taken place since I worked there.

However, I was frozen in my chair. I could say that I didn't speak up because I was nervous, which I was, but that wasn't the real reason. I didn't say what I really thought because, like everybody else on that panel, I was taking them for a ride. Because I wanted a Mercedes Benz, just like Clayton's. I could have told them that the Sound—as they knew it—was as good as dead, which is what I truly thought. Instead, I trembled nervously and lay my head on the table. I missed a chance to be real. I lied.

More Spills and Thrills

March 1977—
Columbia, South Carolina

For some time, I had been thinking about starting a scientific consulting business and eventually leaving the university. It all came together in my mind during that first project in Kuwait in November 1976, and, sure enough, we incorporated Research Planning Institute (RPI) in March 1977. I was the principal shareholder, with JM, Erich, Tim Kana, Bill Campbell, a local politician, and Jim Davis, my personal accountant who was a professor in the USC-East business school, as minor shareholders. We didn't do a whole lot of business in the first year. The Kuwait project was the only one we had for a while, but in the second year, we accrued several hundred thousand dollars in sales. Although I resigned the department head job in the summer of 1977, I did a balancing act between teaching and business for the first few years, resigning from the university to part-time status in 1981.

RPI carried out a contract for a private firm in Kuwait in October 1977, during which I visited oil facilities along the Gulf coastline of Kuwait and Saudi Arabia. We concluded in our report for the project that about five million gallons of oil were spilled into the Gulf each year by normal tanker operations. At the oil terminal at Amina Al Ahmadi[29] in Kuwait, we stood on a tower looking down at a 228,000-ton tanker.

I looked on in amazement, finally saying to my associate Jacob, "You could play three football games simultaneously on the deck of that thing." I couldn't imagine what a 600,000 tonner would look like, which was what the ship builders were contemplating at that time.

In March of 1978, a major oil spill occurred when the tanker *Amoco Cadiz* ran aground on some rocks near Brest, France, spilling seventy million gallons of crude oil (6.2 times as much as was spilled by the *Exxon Valdez*). By then, we had a contract with NOAA to respond to spills for them, and because the French government had not yet completely organized their own response team, they asked NOAA to help out, and we were off to France, by invitation this time. Working with Erich, Craig

Shipp, and several others, we mapped the distribution of the spilled oil, which impacted two hundred miles of shoreline, surveyed beach stations, and generally helped in any way we could. Erich eventually wrote his dissertation emphasizing the combined impacts of that spill and the *Urquiola* spill.

The NOAA team was run by John Robinson, an engineer and inspirational manager, who had led the program that funded our Alaska mapping projects. John was assisted ably by Dave Kennedy, a pilot and jack of all trades, who now runs the NOAA HAZMAT program we work for. The dispersal of the *Amoco Cadiz* oil by currents and wind action was analyzed by NOAA's Jerry Galt, a fiery red-haired oceanographer who, by now, probably has responded to more major oil spills than any other scientist on the face of the earth, with the possible exception of JM.

Just like the coastline oiled during the *Urquiola* spill in Spain, the area affected by the *Amoco Cadiz* spill contained a wide variety of coastal habitats. Consequently, we were able to confirm, with only slight modifications, many of the observations made in Spain on the impacts of oil on different shoreline types.

On a morning several days after the spill, Dave Kennedy and I walked out on a wide intertidal sandflat that had shown no effects of the oil the day before. However, the dispersed oil plume coming in the following night had killed a tremendous number of animals, mostly razor clams, cockles, and burrowing urchins. These dead invertebrates were piled up in swash rows by the waves, some of which were a foot or so high. I estimated that two and one-half million dead animals were present on the surface of that flat, which was only about a thousand yards long and five hundred yards wide. However, few traces were left of the oil itself. This observation led us to formulate our fourth rule of oil-spill behavior: *initial impacts on exposed tidal flats are usually severe, but oil does not normally penetrate the substrate, and it is quickly lifted off the lower intertidal zone by the rising tide.*

A diving team, composed of Tim Kana, JM, Jerry Sexton, Erich, Dan Domeracki, and yours truly, returned to France in mid-summer to determine how much oil had accumulated in the offshore zone. We thought this oil originally floated against the shore, picking up sediment which caused it to sink. It was later transported offshore by bottom currents. The French biologists working with us concluded that the bottom oil had impacted the food sources for some of the important bottom-feeding commercial fish species in that area.

The *Amoco Cadiz* spill was a valuable learning experience, from both the standpoints of scientific knowledge and of the logistics required to organize an efficient method of attacking such a large problem in the field. I believe it was after this spill that Jerry Galt, of NOAA, compared the USC/RPI field teams to a pack of hunting dogs, such as beagles or fox hounds, on short leashes in a single bunch ready to be unleashed at the spill site. Once released, they scattered to the wind, seeking out every possible trace of oil and critical piece of evidence. I liked that description.

I have already alluded to our Christmas of 1978 exploits at the *Peck Slip* spill in Puerto Rico in the Foreword to this book. The USC/RPI representatives at that spill consisted of myself, JM, and Geoff Scott, a biologist. Will Davis, a biologist with an EPA lab in South Carolina, also worked in the field with our group. NOAA rented a house where we stayed with the rest of the NOAA team. For some reason, a number of bureaucrats from Washington showed up, and Dave Kennedy had to entertain them, as it were. We spent most of our time away from them in the field.

One night we returned from a survey about 8:30 P.M., and Dave was running a briefing in the living room of the house, with all the bureaucrats in attendance. Detouring by the meeting, I headed to the kitchen and started scrounging around for a peanut-butter sandwich. I wasn't ready for such a meeting on an empty stomach.

I could hear the proceedings as I ate the sandwich alone in the kitchen. One of the bureaucrats was explaining to the group that a major problem to be anticipated at this spill was that the beaches would become oiled and slick and suddenly slide off into the ocean. Once that happened, the beaches would be lost forever, and the tourist industry in Puerto Rico would be in big trouble.

This discussion went on for quite a while, until finally Dave came into the kitchen and said, "You've got to go in there and listen to that discussion and help me out. They are worried about the beach sliding into the ocean."

"I know. I heard."

I laid down the partially consumed sandwich, walked into the room and interrupted them, making the following pronouncement: "This talk about the beach sliding into the ocean is the stupidest concept I have ever heard. Know why? Because I have profiled a thousand beaches all over the world, and I have never seen any beaches slide off into the ocean, oiled or otherwise. Furthermore, I personally guarantee you that none of you will ever see that happen either. Why don't you talk about the real problems out there, such as the oiled mangroves or the sediment-laden oil that is washing off the beaches out into the grass beds?" Then I went back and finished my peanut-butter sandwich.

These kinds of inane discussions go on at oil spills all the time. At lunch today (the day I am writing this), JM and I were discussing a committee that the National Research Council was organizing to address the problem of heavy oils (that is, oils that don't float). She was invited to serve on this committee, which is an honor.

I said, "Why have a meeting? You already know how the oil behaves in the water, and there is no equipment or technology available to deal with it. Period."

She said, "No wonder you are never invited to serve on any of these committees."

Still the same story, eighteen years later.

The major discovery for us at the Puerto Rico spill was that some large patches of mangroves were killed by the oil. In many ways, the impacts on the mangroves were similar to those we had observed in salt marshes, so we gave mangroves the same high rank of sensitivity to oil-spill impacts that we had awarded salt marshes. That completed the cycle of oiled coastal habitats. Our ESI habitat rankings based on the spills we had studied up through the end of 1978 "have stood the test of time" (a direct quote by John Robinson a couple of years ago). The ESI mapping concept was off and running; we just had not, as yet, experienced an oil spill along a coastline that had been previously mapped.

Meet the Press

Very often there are stories put in the newspaper that I am to be hanged. I don't want that anymore. When a man tries to do right, such stories ought not to be put in the newspapers.

Goyathlay (Geronimo)

Sunday, 18 March 1979—
Los Angeles

The International Oil Spill Conference was held in Los Angeles the week of 18 March 1979. RPI had an exhibition booth at that conference, the primary trade show in the world for those aspiring to compete in the oil-spill business. We enjoyed the fancy Bonaventure Hotel with its exotic architecture. Erich and I gave back-to-back talks on the *Amoco Cadiz* spill, which garnered a certain amount of praise ("we were both brilliant" is what I wrote in my notes). Everyone was complimenting RPI, the new oil-spill response company, and that was our first head trip of that type. Also, our work on the *Peck Slip* spill was discussed often. This meeting was a breakthrough for RPI, and we concluded that we were finally serious players in the oil-spill game.

By the time of the conference, JM and I were living in a two-story English Tudor house built around 1910 that was located about two blocks from the university. It was one of the oldest houses in that part of town, which was burned to the ground by General Sherman's troops in February 1865. JM and I had bought the house a little over a year earlier from a builder, who left town before he finished renovating it. The old house had such an elegant interior that I was inspired to take an interior decorating class at night school, the only male in the class, hoping to adequately finish what the builder had started with such panache. Just in case you haven't taken an interior design course, *panache* is a word designers use to mean enthusiastically artistic or with a flamboyant style.

In April, I gave a talk entitled "Major Oil Spills: Shoreline Pollution and Contingency Planning" at the Fourth Conference on Pollution and the Environment in Kuwait. I showed 156 slides in my fifty-minute presentation and no doubt talked too fast for many people to understand me. Nonetheless, you always have a great audience in Kuwait.

On 2 July 1979, a small audience gathered at McKissick Museum of USC-East to view the works of the Well-Traveled Cameras of Eadweard Muybridge and Miles Hayes. Yep, those cameras had been all the way down to the Assembly Street Pawn Shop and back. My daughters, Joy and Mya, helped me frame the pictures for that exhibit, which didn't exactly take the world by storm.

After that, JM and I went on a dull ten days of sensitivity mapping in Puget Sound. While we were there, John Robinson called to say that oil from a blowout of an offshore oil well, the Ixtoc I, in the Gulf of Mexico off the Yucatan Peninsula was now heading for the Texas coast. NOAA was already mobilizing in Corpus Christi, and he wanted us in Texas as part of the NOAA effort. Before it was capped, many weeks later, the Ixtoc I blowout became the largest oil spill in history (140 million gallons; thirteen times as big as the *Exxon Valdez* spill). To this date, that spill has been surpassed by only the Gulf War spills in total volume spilled.

JM and I arrived in Corpus Christi on Sunday night, 15 July, where we were met by the first wave of our team. All of our technical people at RPI, and many university students, were involved in our response to that spill.

I met with John Robinson, and he told me that, sure enough, the oil was headed for Texas, and that Jerry Galt had predicted that the oil would be crossing by the mouth of the Rio Grande and striking the Texas coast by early August.

"Now is the time to see if your vulnerability index works. How long will it take you and your people to map the coast between the Rio Grande and Aransas Pass?" he asked. That stretch of shoreline was about 150 miles long.

"I think about ten days," I responded.

He laughed. Then he showed me a huge stack of pink telephone messages, all unanswered. He told me there were 124 of them.

"This is it. Our program makes it or breaks it here." These were the words of wisdom he left me with to take back to our waiting field crew.

Next day, we took two jeeps and drove the outer beaches of Padre Island all the way down to the Port Mansfield jetties, about two-thirds of the way to the south end of the island. We ran some profiles and worked out a strategy on how we would do the mapping. We would also have two helicopters just for our use.

It felt strange to be back in my dissertation area. I had worked there off and on for four years between 1959 and 1962. What surprised me most was how beautiful the beach and dunes were. We frolicked in the sun on the beach, excited about the prospect of the job that lay ahead. We stayed late down by the jetties, driving most of the way back through the soft sand and shells in the dark. We had a dinner of Whataburgers on the hoods of the jeeps at 11:30 P.M. This was just the beginning of a summer of long days and nights for that crew.

Toward the end of the mapping survey, we got little sleep, flying in the helicopters all day and entering data on the maps at night. Tuesday, 23 July, was the last day before the maps had to be presented to the local Regional Response Team (RRT) and the Coast Guard. I stayed up until 5:00 A.M. I had to sleep a little bit before the big presentation day, and Dan Domeracki and JM stayed up the whole night coloring the maps.

The most important day ever for the ESI concept (we called it the "vulnerability index" then) was 24 July 1979. I had to present it to the RRT at 10:00 A.M., the Coast Guard at noon, and at a news conference at 2:00 P.M. The maps showed the coastal habitats of the southern third of the Texas coast, ranked according to their sensitivity to oil-spill impacts. This ranking allowed decisions to be made about the allocation of manpower and equipment for the cleanup. We also showed the location of all the areas of major bird congregation and noted that booming strategies would have to be deployed for the three major tidal inlets. Of critical importance were any areas where the oil could get from offshore into the more generally sensitive lagoonal areas. In addition to the three inlets, these areas included washover zones (during extra high tides) and minor tidal inlets, areas that could be closed off with bulldozers, if necessary, which is what eventually happened.

About 8:30 that morning, I met with the NOAA group, assuring John Robinson that I didn't expect any problems with the presentation to the RRT, brimming with confidence as I showed him the finished maps.

John asked, "What if sometime you made a presentation, and they didn't think it was the greatest thing since sliced bread, what then?"

"Would you mind repeating that question?" I replied instantly.

Surely he hadn't asked what I thought he had. These maps were the ticket!

"I think you have your answer, John," Nancy Maynard laughed. She had been listening to the pre-presentation.

As I had expected, the RRT presentation was no problem. After all, this was my dissertation area. But the Coast Guard meeting was another story. Actually, we only met with the federal on-scene coordinator, a Coast Guard captain. It was a dress rehearsal for the news conference at 2:00 P.M. The captain asked Jerry Galt what he was going to say to the press, and Jerry responded, "I'm going to tell them that the oil will be on the Texas beaches in ten days, on 4 August."

"I don't want you to say anything at the press conference!" was the captain's response to Jerry's confident prediction. Incidentally, Jerry's prediction of when the oil would get to Texas was right on the money. I always thought that was a great opportunity lost for NOAA to show off its spill-response skills.

The captain went around the table, censoring almost everything the scientists and other types of responders were going to say. He didn't censor our maps, but it was pretty obvious that I would be up late in the program.

As best I can recall, one result of that meeting was an agreement with our recommendation that protection strategies had to be focused at the major tidal inlets. There were only three permanent inlets—Brazos Santiago Pass, located just north of the mouth of the Rio Grande, the jettied artificial inlet at Port Mansfield, and Aransas Pass, a large natural inlet at the north end of Corpus Christi Bay and south end of Aransas Bay. Therefore, Tim Kana and his assistants from RPI worked with the U.S. Coast Guard strike team to develop booming strategies for those three inlets, which meant taking current measurements in each inlet for many hours.

At last 2:00 P.M. came and with it my first opportunity ever to meet the big-time press. About seven of us, including John Robinson and Jerry Galt, sat at a table on either side of the captain, who obviously meant to have firm control of the meeting. About fifteen microphones, labeled CBS, NCB, ABC, and so on, were right in front of us. Five or six TV cameras were running, representing all the major networks and several local Texas stations. The lights almost blinded me. The room was packed with reporters.

The captain started the meeting by talking for several minutes. I wish I could recall exactly what he said, but I was so dumbfounded by the speech that I am not even sure that I heard it right. What he did say was a bunch of sentences strung together, taken from a source completely unknown to me. I had not heard any of what he said in the RRT meeting, nor in the Coast Guard briefing, nor in any of the numerous science meetings that were held every day. In short, it was media speak, I guess.

After about five minutes or so of his talk, NBC and ABC were gone, and CBS followed shortly thereafter. In fact, he didn't stop talking until all of the TV cameras were turned off and removed, which was only ten or fifteen minutes from when he started. By the time I began talking about the maps, only one reporter was left, the science reporter for the *Corpus Christi Caller*, who wrote a fairly thoughtful article about the spill. Unfortunately, it was published far back from the front page, and I doubt if many people read it.

I was back in my motel room that afternoon before the others returned from the field, so I flipped on the early evening news to see what, if anything, they would show about the news conference. The oil-spill piece started with a long-distance shot of a reporter walking up the beach on Mustang Island. Then the camera zoomed in, and the reporter said, "There's good news and there's bad news."

Then he proceeded to say a number of things that I had not heard at the RRT meeting, nor at the science meetings, nor at any of the Coast Guard meetings. He also totally ignored everything the captain had said in his speech at the news conference. This was all new material. Where did it come from? Then I thought, "Guess that old captain knew what he was doing after all."

Once the oil came ashore, NOAA began tracking the oil and contributed significantly to the deliberations on cleanup options. The booming strategies were successful, and little oil passed through the inlets. One of our jobs was to do field surveys to describe the oil distribution, a process now called SCAT (shoreline cleanup assessment team) surveys. We were also commissioned to continue the sensitivity mapping farther up the Texas coast. I thought the RPI field team was fairly well organized by then. We rented a small office in Corpus, which we maintained for several months.

Meanwhile, I was flying back and forth between Corpus Christi and Columbia, where I tried to fulfill my duties as university professor and principal manager for RPI. During one of the home visits, I was interviewed by a reporter from the local paper in Columbia, *The State,* and on 8 August a picture of me pointing at a map of south Texas appeared on page one of section B. The caption under the picture stated that "members of his Coastal Research Division predicted" that the oil would hit the Texas coast on 4 August, a totally erroneous assertion. Of course, Jerry Galt of NOAA made that prediction, as I carefully told the reporter in the interview, furnishing printed documents and other materials. But, in fact, that article in *The State* wasn't too bad as newspaper articles go, only about one-third of it was wrong.

The article reported that I said "Padre Island provides an excellent barrier for the coastline, but there are creeks and streams that could carry the oil further inland." True, I did say that the outer beaches were ranked a three on our sensitivity scale, which meant that those beaches would be a better place for the oil to go than inside the *tidal inlets* where the more sensitive habitats were. Why this idiot didn't call the openings to the bays and lagoons "tidal inlets" and not "creeks and streams" is beyond me. I'm sure all of the locals who bothered to read that article knew what a tidal inlet was. South Carolina is famous for them. One town on the coast is even named Murrells Inlet.

I went down to Corpus a couple more times in August, and during one of the trips, I was in one of our two helicopters on a beach inspection tour. We landed for gas, and I received an emergency phone call from John Robinson.

"Where are you? CBS is here, David Dick, and he wants to interview the beach survey team. He and his camera crew are in a helicopter headed in your direction. Are you in any position to do a demonstration for him? He's in a hurry. Has a deadline to meet."

"Well, right now we are getting fuel over here in Aransas Pass, but we will be going over to St. Joe Island shortly. The other team is already on the ground at Harbor Island, running a transect, which should make for some decent filming. There is some oil there, and Geoff is looking at the biological effects. I could direct Dick over to them. Why don't you send them on this way? The other team is only a few miles from here."

"Okay. Can you trust those guys to act normal?"

"Hard to say. I'll try to give them a call."

That was a good question John asked, because that crew had worked every day for weeks, their only relief being comical interludes of play acting. Geoff Scott, the Ph.D. biologist on the team who we all called "Dr. Geoff," once led a rock-and-roll band that entertained the tourists at Myrtle Beach, so he was usually the star of the show.

The CBS crew joined us as we were refueling, and we guided them to Harbor Island. I was as worried as John was about how the field team would welcome CBS, and I did try to call them when we got into the air, but there was no answer. Once they came into view, I could see why. Their helicopter was parked on a small dry mound, but the crew, including the pilot, had walked a ways from the chopper through the marsh, which had puddles of water in it, to the location where the most oil was. When we passed over, they were in the middle of a survey, with a tape stretched across the marsh. I asked our pilot to tell the pilot for the CBS crew to land by the helicopter on the ground, and then I waved good-bye to David Dick.

In addition to Dr. Geoff, this survey team included two geologists, Dan Domeracki, otherwise known as "Demo Dan" because of his tendency to wreck cars (I believe the tally was seventeen at that time), and Ray Kaczorowski, who we just called "Kacz" for short. Obviously, I was not on the ground to observe what actually happened, but according to Demo Dan, something akin to the following events ensued.

When the helicopter landed, the survey team didn't look up, thinking it was me and that I would walk on over to them. When the CBS guys noted that they were being ignored, one of them yelled out, "Hey, we're with CBS and want to do an interview on your work with the oil spill."

"Come on in, the water's fine," Kacz responded.

David Dick said, "Hey listen guys, it's wet over there. I just bought this brand new pair of cowboy boots. I can't wade in that mud. How about coming over here by this high ground, there is a little marsh here too, and just simulate that survey you are doing for us so we can capture it on film."

Demo Dan muttered something under his breath about this waste of time but responded politely, "Okay, just give us a minute to mark the position we have reached."

Then they went splashing through the puddles and quickly started acting out another survey, with two profile rods and a measuring tape. The camera was running, and David Dick was talking in the background.

Suddenly, Dr. Geoff reached down and picked up something, walked over and held it up in front of the camera, and started gesticulating wildly, saying: "Ohhhh, Mein Gottt! This is a dead crab! Kacz, Demo, come quick. Look at this poor little dead crab that that awful old oil spill has killed! Ohhh, Mein Gottt! Just look at it, the poor thing. What are we going to do. Ohhh noooo!" The camera man was still firing away, and Dr. Geoff kept on waving his arms and bemoaning the loss of the crab.

"Stop the camera," David Dick said, "Okay fellas, now that you've had your fun, get on back out there with those sticks and tape and you, the blonde comic [Dr. Geoff], act like you are doing something half way intelligent for a change."

"Sure," said Dr. Geoff.

Demo could tell that Dr. Geoff's traveling show was about to start again, so he said, "Listen Geoff, let's get this thing over with. We have four more surveys to do today." And so they did.

Either that evening or the next, I was watching old Walter Cronkite do the evening news when he made a brief report on the cleanup of the largest oil spill in history. Then he called on David Dick at the scene in Corpus Christi, who started talking against a background of Demo Dan, Kacz, and Dr. Geoff running a profile across the marsh. Then they switched to a close-up of our eminent coastal biologist, Dr. Geoff, talking (seriously, this time) about the important science being accomplished at that spill. In the lower right corner of the image appeared the words "Dr. Geoffrey Scott, Research Planning Institute, Inc., Columbia, South Carolina." That was the first time that RPI would make the national news, until ten years later when JM became famous at the *Exxon Valdez* spill.

JM and I both came home for the start of the school year in early September. One day the local TV station interviewed me to fill them in on the spill for their evening news coverage. That night, we watched the news while we gobbled down a quick meal before going back to work. My interview was edited down to a thirty-second sound bite.

"That was short," quipped JM.

"Yep," I responded, as I got up to answer the phone.

"Are you that red-haired Miles Hayes that I used to date down in Wichita Falls, Texas? Saw you on the TV, and you look a little like him."

"Oh. No ma'am. I used to live in Texas, but I've never even been to Wichita Falls," I said as I hung up the phone. JM gave me a funny look and asked me who it was.

"One of my fans," I laughed. Then the phone rang again.

"Is this the Miles Hayes who was just on the TV?"

"Yes, what can I do for you?"

He introduced himself, saying he was calling from Cayce, South Carolina, which is just across the river from Columbia, three or four miles from where I sat trying to finish my dinner.

And then he asked, "Did you ever repair a Volkswagen engine?"

"Well, no. I'm not much of a mechanic, actually."

He could hardly contain the disdain in his voice for someone so ignorant that they didn't even know how to repair a Volkswagen engine, but he continued, asking, "Ever been to Mexico? Do you know what a sombrero is?"

"Well, yes. In fact I have been to Mexico many times."

"Good. Now I gather from your short speech on the TV that the oil well off of Mexico is still spewing oil into the ocean. Well, I know how to stop it."

"Really, a lot of the world's best petroleum engineers have been trying to figure that out for many weeks."

"Now that I know you understand what a sombrero is, that will make my explanation much easier. What you have to do is build a giant, metal sombrero and tow it offshore, turn it upside down, and put it down over the well. The oil will then accumulate in the top of the sombrero where it can be pumped out." He continued with more details that I don't remember.

Finally, I interrupted his lengthy explanation by saying, "You know this is a Mexican problem. As far as I know, they have not asked for help of any kind from the United States with regard to this spill. Pemex, the national oil company of Mexico, is handling it."

"Yeah, I know, and I have been trying to call them, but I can't get anybody to listen to me. You know anybody I can call?"

I remembered one of the graduate students at UT when I was there was on leave from Pemex, and by the summer of 1979, he was pretty high in the ranks of the Pemex management. I didn't particularly like that student, so I gave his name to my mechanical genius caller from Cayce, South Carolina.

"Wow! That's great. Thanks a lot!"

JM and I laughed and joked through the rest of the meal and for the next few days about the sombrero hypothesis.

On Thursday, 13 September, I began the morning with a meeting with the accounting group at RPI, and they told me we had passed the half-million mark in sales for the year and would probably make our goal of eight hundred thousand dollars.

"Sounds good, but is it really true?" I thought as I was boarding the plane to go back to Corpus Christi, where I just made the tail end of the science meeting at the new NOAA quarters, a condo out on Padre Island.

After the oil came onshore, a difference of opinion had developed between NOAA and the state of Texas about what areas to clean up first. NOAA's assertion, based in part on what we had told them and the mapping results, was that the outer beaches, which were mostly hard-packed, fine-grained sand, would clean up naturally. I'm not sure what his official status was, but our old hero from the *Metula* spill in Chile, Roy Hann, an engineering professor at Texas A&M, was the leading advocate for an immediate cleanup effort. Frustrated because his advice was not being taken, Roy turned to the national media, eventually being featured in *People* magazine.

Anyway, Roy was at the NOAA science meeting in Corpus, and we rode together to the Coast Guard meeting that followed. Roy told me that Erich, who was running the beach survey part of the project for RPI, was not representing me properly at the spill. Erich was too pushy and didn't listen to the advice of others, namely himself, I presume.

"Well, you have to remember, he is a New Yorker," was my response.

Roy explained to me all the moves he had made, including the spread in *People* magazine, because the state of Texas was not getting a fair shake.

Next day, we inspected the beach from the Rio Grande to Big Shell, a shelly beach exactly in the middle of Padre Island. The Big Shell area still contained some tar balls and subsurface oil, but the rest of the beaches had been cleaned by the waves of Hurricane Frederick and the ensuing north winds. I thought that this confirmed our predictions about the beach cleaning up naturally. Little effort had been expended up to that time to mechanically clean those beaches.

I made a speech that night at the NOAA science meeting, trying to offset some of the things that Roy had said in his speech before mine. I became a little emotional, explaining that everything was behaving just as we had predicted, so what was all the fuss about?

Nancy Maynard read a poem about two generals fighting a war. I guess Roy was one general, and I was the other—a pretty one-sided fight.

I gave a little calmer version of that speech at the Coast Guard meeting that followed. I also told them that the beach was behaving as expected, that they had been classed as threes on the vulnerability index. I pointed out that the remaining oil at Big Shell was no surprise because those beaches were sixes on the scale. The beach sediment was composed of gravel-sized shell, and oil tends to penetrate and become buried on such beaches.

Next morning at breakfast, I was reading the *Corpus Christi Caller* and noticed an article which stated that Brown and Root, a huge engineering firm in Houston, was going to build a "sombrero" to trap the oil escaping from the runaway well. I never did find out if the Cayce, South Carolina, caller had anything to do with that attempt, but, rest easy, it didn't work.

After breakfast, we went back down to Big Shell and went diving in the surf looking for oil. We found some tar balls being rolled along the bottom by the wave action. Later, Erich and his crew would discover seventeen patches of oil called *tar mats* at the toe of the beach along central Padre Island. Our team never did find an adequate explanation for those features, which stayed around for several years. I always felt bad about that because, otherwise, our response to that spill had been right on the mark.

At 4:00 P.M. on that same Saturday afternoon, 15 September 1979, I was sitting in a jet plane in the Corpus Christi airport holding a one-way ticket to Columbia, South Carolina. "Guess the party's over," I thought.

And it was. I would personally make no more trips to Texas for that spill, although our RPI team maintained a presence there for some time afterward. I was beginning to sink into what Jerry Galt termed the *post-spill grand blue funk.* When that happens, you feel despondent for days. A supercharge of adrenaline is no longer pumping through your veins. You start to wonder about the answer to the primal question again. You may not fully recover from that slump until that next fateful phone call comes that begins with the ominous words "this may be the big one!"

In December, I went to Manila for a workshop on coastal area development and management sponsored by the United Nations. I was pushing the "vulnerability index," but that marketing effort was about ten years ahead of its time. I left there at 6:00 P.M. on 8 December in a sunset over Luzon, almost thirty-eight years to the hour after the Japanese had flown down the valley and bombed Clark Field into oblivion.

I stopped in Honolulu to visit my two daughters. Barbara Anne's new husband, John, was on sabbatical leave at the University of Hawaii. The next day, Joy, Mya, and I drove up to the crest of Roundtop Mountain where we stood looking down on top of the condos of Honolulu, Waikiki Beach, and Pearl Harbor. I hugged them both for a while as we looked down. It was 11:15 A.M., 9 December 1979. It felt like an historic moment.

On the way down the mountain, I said that I was sorry that my visit had been a little short. "Extremely short," Mya said, as she rode in the front seat beside me.

On 17 December, Erich and I left Denver and drove through the brown cloud on over to Boulder, where the NOAA HAZMAT office was located. We made plans with John Robinson and his associates for more oil-spill response and sensitivity mapping in the future.

Later, I asked John how RPI was performing for him, and he replied, "I guess I don't have to tell you that you guys saved my ass at that Texas oil spill."

"Glad to be of service," said I.

Round Island

On this Saturday night, I divided my time between a play at the university in which JM's younger brother Jerry had a role and a party given by one of my graduate students, Mohammad Al-Sarawi, who was leaving the university soon with his Ph.D. in hand. At his party, Mohammad wanted me to come outside where we could talk. We stood on the second-floor balcony overlooking the backyard and watched a possum slowly climb up a tree. As we watched, we talked a little bit about the meaning of life and a whole lot about the great things Mohammad would accomplish in the coastal geology arena when he got home to Kuwait. He has stayed faithful to that mission even to this day.

By 1980, the American Association of Petroleum Geologists-sponsored, six-day field seminar for oil-exploration geologists that I was doing on the South Carolina coast, with the help of most of my graduate students, had taken hold, and we were doing up to eight a year. The courses were fairly profitable for the company, but they also offered the opportunity to market the participants, assuming we had something to sell them. In those days, we started the seminar at the coast for five days and flew a squadron of small planes to Columbia on the next-to-last day. On the night of the overflight, we had a cookout and party at our house, with music, dancing, and carrying on. And on the fateful night of 18 March 1980, during the final party for the course, John Horne and I sold Dome Petroleum, of Calgary, Alberta, on hiring us to do a detailed geological study of one of their oil fields. Thus, a new business was born that would eventually dwarf all other RPI activities.

In the early fall of 1980, JM and I bought Wide Awake, a historic ferry landing located on the Intracoastal Waterway south of Charleston. Among other noteworthy historic events, it was the site of a battle during the First Civil War (aka the Revolutionary War). The house, which had been remodeled when we bought it, was originally built in 1790. It sat on a low bluff by the channel overlooking several thousand acres of salt marsh across the waterway. The property we bought contained seven acres on the waterway itself, including four acres of lawn with marvelous flowering plants, such as azaleas and camellias. Wide Awake became our weekend escape during the hectic business-building years of 1980–1983.

We drove down to the coast in the late afternoon of 10 October 1980 to spend our first weekend at Wide Awake. I woke up on Saturday morning and looked out over the marsh, just before the view was partially blocked by a huge sailboat moving down the waterway. I was freaking out, wondering where on earth I was.

Later, I walked outside to take some pictures and noticed a lone, scarlet tiger lily in the middle of the grass beside the house. Several bright yellow butterflies hovered around it, a rare sight. Even more impressive were all the birds in the marsh and in the trees around the house. Almost instantly, JM and I took a strong interest in the birds, buying field guidebooks and starting a bird list. Within a year or two, we were listing over two hundred species a year in South Carolina alone. That first weekend, a bottle-nosed dolphin paid us a visit every morning, performing acrobatics in the waterway by a green buoy marker across from the house.

Toward the end of October, we ran another field seminar on the coast to which we had invited a number of upper-level managers of the oil companies in Calgary. The managers were mostly geologists in those days, not lawyers and accountants as they are now, so they were interested in learning how the world was made.

Thinking back on those days in the field as I rode on an airplane headed for Anchorage on 27 October, I wrote the following in my diary:

I have just spent a week with another field course. The weather was rotten the whole time, but it was fun to be out in the waves and the wind. We started by going to the Santee Delta. It rained. And then it rained and rained some more. Every day thereafter, the schedule had to be changed. One night I showed "A Day Just Like Today," and we had a party at the house on Sunday night. The party was very quiet, only the participants in the seminar came.

By the way, I had fun out there in the rain on Kiawah Island on Thursday. The wind was blowing rain and sand. It reminded me of the "old days" on Plum Island, Massachusetts. A regular storm! Sand blowing over the flats, filling up the trenches before we even got there to view the structures. For a while, I reveled with nature. They didn't quite understand that we were having a conversation with the earth. With the Great Spirit. With the essence of truth!

With clouds and rain and sand—it's 62 degrees below zero at 35,000 ft.—Ha, ha. Drunk again!"

Shortly thereafter, we bought an old abandoned warehouse on the main street of town, one and a half blocks from the state capitol, for eighty-four thousand dollars. The reason for the purchase was to design and build a new office for RPI. JM, Dan Domeracki, and I designed it; our design included polished original heart-of-pine floors, sand-blasted brick walls, and a spectacular spider-shaped heating/air conditioning unit hung from the high ceiling of the main room. The spider was painted bright yellow. Along the center of the room ran a series of round poles that we painted a spectrum of hues of orange to contrast with the blue walls of the open-space offices—opposites on the color wheel, you know. It was glorious; it was beautiful! We moved in for the first day of business on 8 February 1981.

Soon thereafter, I told my dean that I wanted to go part-time at the university. I had places to go—Nigeria, Calgary, Kuwait—and things to do.

The hectic pace continued through February, March, and April. We did manage to sneak a few days off at Wide Awake for our first spring there, with its full blossoming of the grand azalea bushes all around the place. About the month of April I wrote in my diary "Flowers and short courses. Never a moment to stop and think about why it's all necessary."

Sunday, 2 August 1981—
Flying in a Cessna 180 between Anchorage and King Salmon, Alaska

By 1981, I didn't go out on spills any more because of my commitment to growing the business. Erich was the manager of the NOAA contract, and JM also went to many of the spills. However, I did reserve the right to participate on the ESI mapping jobs, particularly if they were in an especially interesting area. Such was the case in August 1981, when I joined JM for a mapping project in Bristol Bay, Alaska, an area I cited often in my lectures as a classic example of a tide-dominated embayment, an area where large tides determine the nature of the coast.

On this particular Sunday, JM and I rode in a Cessna 180 we had chartered from Kenai Air on our way from Anchorage to King Salmon. From there, we flew down the familiar shoreline along the backside of the Alaskan Peninsula, heading for Cold Bay. As we continued west, the low ceiling kept getting lower and lower. We discussed having to land on the beach and spend the night. Then the pilot remembered that a lodge was somewhere in that area. He turned inland, up a stream valley, where he thought it might be. Shortly, we were landing on a gravel airstrip next to the Bear Lake Lodge in the middle of nowhere. Next morning, JM and I got up early and birded along the shoreline of the peaceful glaciated lake by the lodge. We saw several species we had not seen before, including a red-throated loon.

After four days of mapping, we had reached all the way around to Dillingham, where we landed to spend the night on 6 August. At the airport in Dillingham, we caught a cab owned by the R. B. Cab Company. A rather nondescript thirty-year-old was sitting in the back of the van when we got in. The cab was driven by a young man with long blonde hair. There was a radio in the vehicle, which the driver had a difficult time operating. He and whoever he was talking with said "roger" in between every other word. As we headed towards the center of town and our hotel, the driver went in and out of several of the small side roads without ever picking up anybody, meanwhile never missing any of the big mud puddles that he splashed through with a flourish.

Somewhere along the line, the guy in the back said in a loud voice: "Hey! Guess who found a fifty-dollar bill today?" A long period of silence followed. He never told us who, but JM and I asked each other that same question several times over the next few days.

Finally, the blonde pulled into another side road, got out of the van, and ran into a shack. The Fifty-Dollar Bill Man promptly crawled over the back of the seat and drove us to our hotel.

Next morning at the Captain's Table, we had just started to eat breakfast when a big fight broke out in the kitchen. The waitress said, "Another typical day in Dillingham." Pots and pans were banging around and flying out from behind the wall that hid the kitchen. The two combatants were trying to decide who was the boss.

We left Dillingham about 7:00 A.M. and flew to Cape Newenham, and I mapped the coast up to Eek on the Kuskokwim Delta. Unfortunately, the fog was so low at that point we had to quit for the day.

However, we decided to fly back out to Cape Newenham and look for some birds on the beach. We spent two hours there under a beautiful midday sun. We saw several new birds to add to our list, including tufted puffins, black-legged kittiwakes, and white-winged scoters. We walked a long way over the tundra. A beautiful spot!

On the way to Bethel, the next place we would spend the night, we landed on a washover beach on the delta. I had never seen so many birds in one spot. The ground was literally covered with sandpipers.

That night, JM and I discussed the value of doing field work. We concluded that I should keep in touch. I was inspired to do science!

On 9 September, we were in the air by 7:30 A.M. and heading for Nunivak Island. The weather was bad again so we landed to wait out the weather at Toksook, a remote native village on Baird Inlet of the Kuskokwim Delta. We walked around the village and looked at some birds on the beach. The village was clean, being replete with "Don't Litter" signs and all kinds of directions about where the inhabitants should run their three-wheelers. I didn't realize until we took off that no cars were there, not one.

As we made that walking tour of the village and beach, I thought, "It's a little sad to see the demise of a culture like this. Sold out to ski mobiles, three-wheelers, and outboard engines."

Just a few salmon were hanging on the racks down by the beach, and John Peter, some kind of social worker, was on TV advertising cultural training courses for all the villagers.

Finally, we took off and flew across the cloudy waters to Nunivak Island, where we mapped the coast for over two hours. The island was a wild, remote, and spectacular place! That was the real Alaska! One high vertical cliff flanks the east end of the island for tens of miles. Thick lava flows were exposed in the cliff face. Literally millions of birds inhabit the cliffs, mostly kittiwakes. It is pretty hard to ID birds at a hundred miles per hour.

Long stretches of fine-grained sand beaches fill in the gaps between the lava-dominated headlands. Dunes over a hundred feet high back up the beaches in places. Most of the island was totally untouched by humans! No three-wheelers! No garbage dumps! No stench! No kids with B-B guns! No preachers! No teachers! No John Peters! Only sand dunes, both barren and vegetated, hills of grass lying over away from the wind, and big waves pounding the beach with its hexagonal volcanic boulders. And herds of musk ox.

When we headed back over the water to the mainland, we had finished the work with the fixed wing, almost one hundred maps in seven days. We had expected the work to take fourteen. We felt we had done an accurate job, and we were surprised how quickly it went.

About 7:00 P.M., we landed back at Dillingham and caught the R. B. cab into town. The cab was driven by Fifty-Dollar Bill Man. He was getting some complicated directions over the radio by his two amateur-hour associates as to how to make a pickup on the way into town. He couldn't seem to get the directions straight, so he just kept driving in and out of every one of the tiny side roads he came to, for about twenty to thirty minutes.

After he finally made the pickup and we were heading back to town, I said, "Hey, guess who found a fifty-dollar bill today?"

He turned back to us, grinned, and said, "Did you find a fifty-dollar bill? I found one yesterday."

"I wonder who's throwing those fifty-dollar bills all over the place?" I replied.

Then he turned back to the front and said, "Base to 3, I'm on my way in." They tried to explain to him that they were "base" and he was "3", but he never changed the way he said it.

Next day, 10 September, the NOAA helicopter arrived, and we went out to map the Walrus Islands, which are located twenty miles or so off the mainland southwest of Dillingham. First we landed at Round Island, a wildlife sanctuary about one and one-half miles long and a mile wide. Jim Taggart, a doctoral student at UC Santa Clara, and his wife were living on the island conducting studies of the walruses and foxes.

In addition to the walrus haulout, a large bird rookery is on the island.[30] Jim took us up a steep trail to the cliff top to give us a close-up view of the birds. As we neared the top, I slipped back down the slope a few feet, clutching at the damp grass alongside the narrow foot path. I could see a few horned puffins on the water below me and wanted to stop to observe them for a while.

"They're much better up over the next hill," Jim said, urging us on up the slope.

Then, realizing that JM and I were puffing pretty hard, he slowed down and asked, "What does an oil spill do to a bird rookery?"

I was surprised how difficult it was to give a concise answer to such a straightforward question. I didn't answer right away.

"Of course, the birds bring the oil back to the nest," JM responded for me.

After further explanation about the oil matting the birds' feathers, causing hypothermia and death, killing the eggs, and so forth, I added that our shoreline mapping system dealt first with fixed habitats and that mobile organisms, such as birds and fish, were treated separately on the maps. I explained the ESI as best I could through all the puffing and panting. Fortunately, when we reached the summit, our conversation was obscured by the din of the rookery. We lay on the grass and peered over the edge of the vertical rock wall at thousands of nesting horned puffins, kittiwakes, and common murres.

Jim pointed to the other side of the cove to a sand and gravel beach which was covered with hundreds of male walruses, the subject of his dissertation. He explained that Round Island was the only walrus haul-out in "the free world."

It was clear that he thought "his" island was important and that maybe there shouldn't be any oil spills there. Since our mapping was part of an oil-spill contingency plan, I couldn't offer him much encouragement. "Well, they could just not drill for oil here," he said.

We saw a bunch of new birds (for us) from the top of the cliff, and then we reboarded the helicopter and mapped Round Island and the rest of the Walrus Islands, completing our aerial ESI mapping of the Bristol Bay shoreline.

By most estimates, as many as 350,000 birds living in rookeries like the one at Round Island were killed during the *Exxon Valdez* oil spill. Those rookeries were located mostly on the Barren Islands, near the entrance to Cook Inlet. Common murres were particularly vulnerable because of their habit of diving off the rocks into the water to catch fish, plus the fact that they roost in large rafts on the open water surface.

We dropped JM off at the Dillingham airport and picked up Dan Domeracki so we could scout some coring sites. Dan would lead ground surveys of the Bristol Bay area after we left. When I got back to the motel room, JM was lying on the bed, looking at some maps with one eye and with the other at a confrontation on the TV screen. The confrontation had happened a day earlier in some town in Alaska. James Watt, the new secretary of the department of interior, was saying on TV that Alaska should be opened up more for people who fished for sport, hunters, and the common people, and needless to say, for more oil exploration. His adversary in the debate, Governor Hammond, told Watt that he was probably not very well informed on the potential impact of offshore drilling on the billion-dollar salmon fishing industry in Bristol Bay. I remembered the hundreds of fishing boats that we had flown over out in the Bay by King Salmon.

Next day, we went on a grueling three-hour ride back to Anchorage in the Cessna 180, flying in marginal weather. I think I was more concerned about crashing on that flight than on any I had ever taken up to that time. That concern was probably due to the fact that we had just heard that one of the Bear Lake taxi planes had crashed, with a Fish and Game man killed and the pilot seriously injured. We had met two men like them at Bear Lake Lodge, and we wondered if they were the ones.

We flew back in the clouds and the fog, sometimes as low as three hundred feet. We passed by some of our 1976 stations on the west side of Cook Inlet but were pressed for time so we didn't stop.

That night we took a jet to Juneau, where we put in one more day of birding before heading home. During that trip to Alaska, we added forty-seven new birds to our life list (first sightings). The most exciting birding was on the cliffs at Round Island and the sandpipers on the Kuskokwim Delta.

The Board of Uncles

Business had been in a slump for RPI during the first half of 1981, but through the summer and fall, things started to break in our favor. We landed a four-million-dollar project in Nigeria and started our second company, RPI Colorado, Inc., in Boulder in September. December business meetings showed projections of several million dollars in sales for 1982.

The next two years were probably the most hectic of my life, as we grew the Boulder office and an office in Calgary, and chased other business leads.

Early in 1982, the RPI management group kicked around the idea of setting up an outside advisory board because we were growing like topsy, mostly without thought being given to the ultimate consequences of such growth. The question was, who should we invite to serve on the board? I thought I would start at the top, so I wrote a letter to Ted Turner, a South Carolina part-time resident, to see if he would be interested. I received a letter back from Ted, who said he didn't have the time right then but wished us the best of luck.

During our discussions, Erich observed that his Uncle Irwin, an executive with Kaiser Aluminum in San Francisco, would have a lot of good ideas. Tim Kana said his wife, Julie, had a number of successful uncles—in particular, Uncle John who was by then retired but had been the president of the largest bank in South Carolina for many years. Okay, so we had two board members to start with—Uncle John and Uncle Irwin.

JM and I were feeling a little left out and began to wonder if we had any uncles to throw into the pot, so to speak. Although my mother and father combined had a total of seventeen siblings, only seven of them made it into the adult world, a pretty good clue as to how hard it was to grow up in the mountains in the early part of the twentieth century. Of those seven, only three were left who qualified to be my uncle. First was

Uncle Obrey, my dad's older brother, but he was a drunk who may not have made it even as far as the fifth grade as far as I know, so he wasn't much of a candidate. His younger brother Junior was in a hillbilly band, as I have already said, but he was totally without corporate experience.

On my mother's side was my beloved Uncle Luster, one hell of a fisherman and a good gardener, but his business experience as a plumber in a textile mill hardly qualified him for matters of high finance. The only one left was my Uncle Linc, the husband of my mother's sister Lily Mae.

I knew my Uncle Linc and Aunt Lily Mae quite well, having spent a week or two with them each summer as I was growing up. They owned a few acres of land up near the foot of the Blue Ridge and were the proud owners of an awesome pack of hunting beagles. Uncle Linc also worked in the mill—that is, until he fell off a ladder and messed up his knee so bad that they laid him off. He was probably not board material either, but you know what he could do? Man, he could really play the banjo. So I thought that if ever the board was in need of an entertainment break, maybe he and Uncle Junior could get together and perform for us. Unfortunately, I never asked them to, and I wish I had.

Both Uncle Linc and Aunt Lily Mae are dead now, but as we were discussing that redoubtable board of uncles a few days ago, JM remembered the last visit we made to the dog farm about ten years earlier. The thing I remembered most about that visit was the fabulous banjo concert Uncle Linc put on at my request. He was up in his eighties then. The thing JM remembers most about the visit was the flatulent fireworks display put on by the old couple as he played the banjo and she packed a bag full of paperback romance novels for us to take to my mother.

JM's dad was a successful insurance adjuster in Charleston and even graduated from college at Notre Dame. He was also a hero in World War II, figuring out how to beach amphibious landing craft while avoiding getting blown up by the Japanese, unlike the unfortunates who went in before him. His dad was a junk dealer, and JM's mom's dad was a taxi cab driver in New York City. As you can see, JM's gene pool was about as sparse as mine when it came to successful business heritage. When we started looking for uncles in her family, we found only two candidates, Uncle Laurie and Uncle Charley. Uncle Laurie, JM's dad's brother, taught at some upstate New York University, but, Lord knows, we didn't need any more academics on the board. We already had too many.

JM's Uncle Charley was afflicted with the same malady as my Uncle Obrey—that is, he was also a drunk. One time not too long after we bought Wide Awake, we gave a Sunday afternoon reception for JM's dad's second wedding. It was a gala affair, with sailboats carrying Charleston's finest to Wide Awake, cluttering the waterway as the guests filled the grounds around the property. Somebody suggested that I show "Suzanne's Lament" one more time for the assembled guests, which I did.

During the middle of the show, Uncle Charley, who had had one too many by then, stood up and said in a loud voice, "Hey Miles, it would be better if you talked while you show those pictures."

The big day for the first and only meeting of the Board of Uncles finally came on 29 November 1982. Of course, the board consisted of more than uncles. One guy, who had successfully formed a company like ours and quit to go back to the university, came up from Miami, and my old office mate from UT, Murray Felsher, who was a successful bureaucrat in D.C. by then, was there, as were all the insiders, of course.

Uncle Irwin told me that he highly admired what our group had accomplished in such a short time with no capital investment, but that we would do even better if we became "the environmental voice for industry." We debated that idea for some time, with Erich, Uncle Irwin's nephew, being the leading advocate. Somehow I couldn't buy that idea. I wanted to be right, no matter what, and being a paid liar somehow wasn't my goal in life, not that Uncle Irwin was advocating anything of the sort or even that most industrial clients want their consultants to lie for them. This was just my personal perception of such a role at that time. Anyway, Brother Dycus would have been proud of me, I think. Most of the others agreed with me. NOAA had brought us to where we were, and we thought we should stick with them.

Uncle John had a different idea. He was amazed that someone with a beard and from an academic background, no less, could actually make money. He thought it unfortunate that I had no political connections and set about to rectify that situation. He set up a meeting between him, me, and ex-Governor McNair, who then ran the largest and most successful law firm in South Carolina. The meeting took place in the ex-governor's office on the top floor of the tallest building in Columbia. I showed up with the usual armful of marketing material and went about filling the governor in on all the great things RPI was doing to make the world a better place to be. About ten minutes into my usual marketing spiel, I could see that I had completely lost the governor.

So did Uncle John, who stopped me and said, "I see what your problem is, Gov, you can't quite figure out just what it is that these folks do!"

About ten minutes later, I folded up my maps and other materials, excused myself from the meeting, thanking Uncle John profusely. Then I rode the elevator down the eighteen floors, walked into the street and then down the hill two and a half blocks to our office, never again to venture into the world of high politics in the grand state of South Carolina.

Next day, Roger Reed, the comptroller for the company, gave me the disquieting news that the Columbia office had lost eight thousand dollars in November, one of our first losing months on record. Then Roger said, "But Colorado beat us out; they lost ten thousand dollars!"

We went into a slump that lasted several months. In order to find the cash to bail us out, JM and I sold Wide Awake. That was the downside. The upside was that I had a good excuse to not convene the Board of Uncles anymore because we couldn't afford it.

After that small downturn in early 1983, we righted ourselves and experienced a continued spurt of growth that included forming new companies in Calgary, Alberta; Port Harcourt, Nigeria; Austin, Texas; and San Francisco, California to go with the two original companies in Columbia and Boulder. JM and I, though missing paychecks for months at a time on occasion, were doing well, becoming paper millionaires and owning a "real" plantation on the coast called "Prospect Hill," the top of a mountain in East Tennessee, and several other pieces of real estate. By the time our business crashed, the company was zeroing in on ten million dollars in annual sales and had almost two hundred employees.

I wasn't doing much in the oil-spill business while the company was in this growth mode. I was flying around in circles playing CEO. Erich ran the NOAA contract for a few years, but in late 1984, he and Tim Kana decided to spin off their own company. JM has run the NOAA contract from then to this day. The Columbia office continued as the environmental wing of the company, with its sales holding steady at about thirty percent of our total companywide sales throughout the growth of the energy sector.

Although many signs of the coming decline of the oil-and-gas industry in the United States were evident in the mid-1980s, we were too undercapitalized to diversify into the international oil-exploration business in time to compensate for the disintegration of the domestic industry. When the price of a barrel of oil dropped to under ten dollars in early 1986, an almost two-hundred-percent drop in a few short months, oil centers like Denver and Houston went into a major catharsis the likes of which had not been seen there since the Great Depression. Just like most of the other companies in the so-called service industry for the oil-exploration business, we struggled on through 1986 while staring at the prospect of bankruptcy every day. The company survived 1986, but the energy program was collapsed back to the Boulder office. This meant that yours truly and Dan Domeracki, who was the chief operations officer of the newly formed holding company, had to travel around the country to the individual offices and close them down, firing the entire staff in some cases.

Things got a little better in 1987; we were making pretty good sales on the energy side, and as usual, the environmental side held steady. But we were carrying a tremendous debt the whole time. The year 1988 started real strong, and our staff was back up to over sixty people, with the Boulder office cranking out several major reports. However, the price of oil dropped again at the end of 1988, and we didn't get the usual year-end sales.

We started 1989 with a major cash shortage, and partly because of an accounting snafu at the corporate level (we had hired a new chief financial officer by then), the first few months of 1989 became the hardest of all with respect to the survival of the business. By then, JM and I were essentially out of resources of any value, so we couldn't help out as we had before. The plantation and the Tennessee mountaintop were soon to go, and our personal net worth was a negative number, a big one at that. At mid-year, a company in Calgary made a substantial offer to buy the energy side of the company, but I never really believed it would happen, and as expected, it never did.

On 11 May 1989, I was riding an airplane looking down on Shiprock in New Mexico. After we passed it, I wrote in my diary, "What do I really think? RPI will live. The dream will live. We may have to sell the energy division, but somehow we will survive. God only knows how. Nonetheless, it will never be quite the same again after this." And that's the way it has worked out.

PART 6

SCIENCE,

PAID LIARS, AND

VIDEOTAPES

When I was going around the world, all were asking for Cochise. Now he is here—you see him and you hear him, are you glad? . . . I do not wish to hide anything from you nor have you hide anything from me. I will not lie to you, do not lie to me.

Cochise

The Day the Music Died

Friday, 24 March 1989 —
Prince William Sound, Alaska

On this day, "the tanker *Exxon Valdez* grounded on Bligh Reef in Alaska's Prince William Sound, rupturing its hull and spilling more than ten million gallons of Prudhoe Bay crude oil into a remote, scenic, and biologically productive body of water. It was the largest oil spill in U.S. waters. In the weeks and months that followed, the oil spread over a wide area in Prince William Sound and beyond, resulting in an unprecedented response and cleanup."

These words make up the introductory paragraph for a NOAA report on the *Exxon Valdez* oil spill written by Gary Shigenaka, a NOAA biologist, in the fall of 1995. JM and I were co-authors of this report, which synthesized five years of the NOAA-sponsored biological and geomorphological research on the effects of the spill.

As you may recall, we were not having a very good time at RPI in late March 1989. The day before the wreck of the *Exxon Valdez*, I was in New Jersey giving a synthesis talk on the barrier islands of the southeast at the regional meeting of the Geological Society of America, rushing back to Columbia late that same night. One never knew when the next crisis might develop. Dave Kennedy of NOAA called our office early the next morning, a Friday, forewarning us that "this was the big one" and we should be ready to go. And, sure enough, JM and Jeff Dahlin, one of the RPI biologists, were off to Prince William Sound by late afternoon of that same day.

The tanker wreck occurred almost twenty-five years to the day after the Good Friday earthquake of 1964. That earthquake destroyed parts of Anchorage and generated a *tsunami* (tidal wave) that wiped out the coastal village of Valdez, the terminus of the pipeline. It also raised parts of the Sound out of the water as much as thirty-five feet. Most of the shoreline in the Sound impacted by the spill was raised several feet during the earthquake, creating some unusual types of gravel beaches. The epicenter of that earthquake was located in the same general area of the Sound as Bligh Reef, the rocks the tanker ran aground on.

I spent that weekend alone at the Bluff, but I was thinking about the spill quite a bit. I remembered the fateful day of the meeting in Cordova in April 1977, when I just sat there, biting my tongue, instead of warning the locals about the spill that I knew would occur in the Sound some day. Now it had happened.

The following Monday, I went to Boulder, continuing the saga of trying to keep the company afloat with the business downturn that had occurred at the end of 1988. On Tuesday morning, I asked the members of the board to sign a resolution that our lawyers could file either Chapter 7 or Chapter 11 proceedings if it ever seemed prudent to do so. The motion was signed with only limited discussion. Filing was not what I had in mind (and we never did), but I wanted to be ready, just in case.

During our usual evening phone call on that Tuesday, JM told me that she had been on national TV, had her picture on the front page of the *New York Times,* and could be heard on NPR and read about in the *Boston Globe.* I told her that I was trying my best to stay out of the newspapers.

I didn't visit the spill site once in the first year. But I guess you could say that I was there in spirit because JM and two of my ex-graduate students, Erich and Ed Owens, played prominent roles in the spill response. In a book entitled *Degrees of Disaster,*[31] Jeff Wheelwright said the following about my friends and associates:

> The three shoreline geomorphologists were independent consultants. HAZMAT contracted Jacqueline Michel, a longtime NOAA advisor. DEC engaged Erich Gundlach. Exxon's coastal specialist was Edward Owens. Drs. Michel, Gundlach, and Owens knew each other well.
>
> "We are the big three of oil spill geomorphology," said Michel. Gundlach said, "I'd say we are the only three."
>
> "In the 70's, Owens and later Gundlach earned doctoral degrees at the University of South Carolina under a coastal geomorphologist named Miles Hayes. Michel was not a graduate student of Hayes — but she married him and now they run a consulting business together.
>
> As Hayes trained them to do, each specialist established fifteen to twenty stations on beaches in the western Sound. They and their assistants collected hydrocarbon samples for analysis, monitored changes in oil cover and penetration and traced the reworking of

beach profiles caused by the elements and the cleanup crews. With a budget of several million dollars and a staff of fifty, Owen's project was the most ambitious of the three.

"A staggering number of observations were made," Owens said. "But we three saw the same things and basically didn't disagree on anything."

"While not exactly the same," Michel put it, "our results are more or less compatible."

Back to Gary's introduction to the five-year *Exxon Valdez* report:

While there is no question that the *Exxon Valdez* spill was an unfortunate, and in some ways, tragic incident, it is also clear that it provided a necessary impetus to reexamine the state of oil spill prevention, response, and cleanup. One result was the passage of the Oil Pollution Act of 1990 by the U.S. Congress. Many states responded in similar fashion by tightening or completely restructuring oversight of oil transportation and production. For NOAA HAZMAT, the *Exxon Valdez* spill was by far the largest incident response ever mustered. In addition, it was a unique opportunity to learn about the long-term effects of oil and cleanup activities in a relatively unaltered setting, and to gain a greater level of understanding that would facilitate a more effective and lower impact response in future incidents.

All of this is true, I suppose, but Gary forgot to mention one thing: when this spill occurred it was also the day the music died. RPI had been responding to spills for eleven years at the time of the *Exxon Valdez* spill. Up until then, going to spills was an adventure unsullied by agendas, legal entanglements, and paid liars unlimited. With all the money and prestige at stake, one had to be careful at the *Exxon Valdez* spill. In many ways, it was the end of the purity of applied science as we had known it. However, I didn't know that yet. I would find out for myself a little over a year later.

Many Rivers Ran Through It

Friday, 2 May 1986—
Foothills Trail, South Carolina

On this particular weekend, JM and I decided to take a three-dayer and hike and camp the Foothills Trail along the base of the Blue Ridge in South Carolina. Some time before that, I was shopping in a bookstore in some airport somewhere, probably in Atlanta, hoping to amuse myself sufficiently on another airplane ride across the United States. A brown-rimmed paperback, which had a blue centerpiece with a big fishhook in the middle of it, caught my eye. I could see the fishhook all the way across the room. On closer inspection, I noticed that in the background was a man standing in a canoe casting with a fly rod.

Fish hooks! A fly rod! This awakened a passion that had died a slow death in the spring creeks of the central Texas hill country back around 1961. I had packed and carried my old fiberglass fly rod around with me from move to move for twenty-five years, without ever once taking it out of its cloth carrying case. That practice was about to end because after reading that book by David James Duncan called *The River Why,*[32] nothing could keep me from fly fishing again as soon as possible.

The first opportunity came on this particular weekend in May 1986 in the South Carolina foothills, when I toted the fiberglass rod along with a big backpack the six miles of the trail to a camping spot on Thompson River. After we set up the tent, I told JM that I would catch dinner, but I didn't have any flies left, only a couple of bare hooks. So I scraped a "stick bait," a caddis fly larvae, off the bottom of a cobble in the stream and after a while was able to catch an eight-inch brown trout, which we ate along with our macaroni and cheese. Not much of a start, but it was all that was needed to get me back into the game. A couple of months later, JM and I went for the Fourth of July weekend to our moun-tain property, Black Branch, in east Tennessee. On the way, we stopped at a fly shop in Charlotte and bought a Green Mountain series Orvis fly rod, the cheapest they had, and used it, plus the old fiberglass rod, on that trip. JM caught the bug, too, and before long, she was outfishing me more often than not.

Although recently popularized by Robert Redford's movie version of a great book, *A River Runs Through It* by Norman Maclean,[33] which I read several years after reading *A River Why*, fly fishing goes back a long way. Most scholars on the subject attribute the earliest literary reference to the sport to a woman, Dame Julianna Burners,[34] who wrote a book entitled *Treatys of Fishing With an Angle* dated 1496. In that book, the author discussed how to tie twelve different trout flies.[35] In the 1600s and beyond, a number of British writers started a proliferation of literature on this, in my opinion, the greatest form of outdoor recreation.

By the time the year 1989 rolled around, I was back into fly fishing in a big way, even tying my own flies, something I had never done before. Fly fishing became my escape from the creditor calls, bank meetings, and the pain of layoffs and downsizing. I went every chance I got that year, when I was spending much of my time in Boulder.

During the long summer days, Dan Domeracki, another student convert of mine to the sport, and I could fish until 9:30 at night, and it was only a twenty-minute drive to several decent trout streams. And, of course, the whole of the Colorado Rockies, with some of the best trout streams in the world, could be reached on the weekend.

According to my fishing notes—yeah, I keep those too—I went fishing seventy-five times in 1989, a record I haven't come close to since. My New Year's resolution is usually to go only sixty-five times a year. The reason I went so much in 1989 was that JM was away most of the time at the *Exxon Valdez* spill site, and I didn't bother to come home from Boulder for the weekend. Even when she would come back to the "lower forty-eight," we often spent our weekends fishing the streams in Colorado. Most of the fishing trips, however, were by myself somewhere around Boulder after work. JM and I had a nice few days in Yellowstone National Park in late July, where we caught large numbers of the big Yellowstone cutthroats. Dan joined us toward the end of that trip, but we got in only one day of fishing after he arrived because of a crisis in the Boulder office.

The one day that stands out in my mind as the very best of the fishing by myself that year was on a Sunday afternoon, 27 August 1989, when I walked up a trail in the Rocky Mountain National Park to fish the Middle St. Vrain River. This was a great day, even the first three hours, when I caught only seven fish. During that period, I was fishing with nymphs along the bottom, and it was very challenging.

The section of the stream I was fishing is lined with large fir trees. Occasionally, I could spot through the limbs of the trees the patches of snow on the highest peaks. It was a beautiful and peaceful late afternoon.

Around 5:15 P.M., I had fished for half an hour with the nymphs without a single strike. For no apparent reason except that I wasn't catching anything on the nymphs, I switched to a dry fly, a No. 18 female Adams. This change was serendipitous because shortly thereafter, a No. 18 gray mayfly starting hatching, and the fish were rising to them. The fly I had just tied onto my leader was a close match for the hatching mayfly. I was still learning the mayflies at that time and still am, as far as that goes, so I could not make a definitive identification of those particular mayfly duns (the hatching adults).

I was fishing a part of the stream where it cut down over the end moraine of the last glacier to exit the valley, and there was a stair step series of small pools where the water dropped over the large boulders left by the ice. The stream split into two channels, with a major pool at the point where they split apart at the upper end of the moraine.

As soon as the hatch started, I hooked a brown trout between eight and twelve inches long in every single one of the pools as I moved up the left channel. After the sixth fish, I broke off the No. 18 female Adams. I only had two female Adams flies left, both No. 16s, a little larger than the one I was using. As I was tying on the first of the larger Adams, I dropped it in the stream, and it was swept away by the current. I broke the other one off on the second fish I hooked after that, so I was out of Adams flies. The closest match I had left was a No. 18 yellow humpy, which worked fine until it was so chewed up it looked like a fur ball.

When I reached the big pool at the top of the moraine, fish were rising all over it. What appeared to be a spinner fall of No. 12 ginger-colored mayflies and a large number of No. 24 midges were in the air, but the fish seemed to be focusing on the hatching No. 18 gray mayflies. I noted on my pocket tape recorder that this was the heaviest insect activity I had seen all year. I caught about five more fish in the big pool fishing from the downstream side, hooking and losing a fifteen-incher. The main current tongue, which was quite strong, was between me and another pocket of rising fish on the other side of the pool. Try as I may, I couldn't get a drag-free float to that area from that position. I then worked my way upstream, walking between and around some boulders, to a position where I could fish downstream and reach the other side of the pool with a collapsing parachute cast, which gave me about a three second drag-free float. (And you thought the part about oil spills was technical!) I caught a couple more in that fashion, finally putting the fish down. In any event, I was out of flies that matched the hatch the fish were interested in. I quit at 6:45 P.M., and according to my notes, I had brought fourteen fish to hand and released them while fishing with the dry flies in that last hour and a half.

By then I was pretty tired, so I scrambled up over the boulders to the trail and started walking back down toward the car. Every so often, I would turn and look back up toward the mountains. The last rays of the sun were shining down the alley between the big trees. The air over the trail was full of the ginger spinners, the No. 18 gray mayflies, and the tiny midges by the thousands, doing cartwheels in the air and glistening in the sunlight. It was quite a sight.

My Father's Grave

I buried him in that beautiful valley of winding waters. I love that land more than all the rest of the world. A man who would not love his father's grave is worse than a wild animal.

Chief Joseph of the Nez Perce, speaking about burying his father, Chief Old Joseph, in the beautiful Wallowa Mountains of Oregon

Friday, 17 November 1989—
Boulder, Colorado

I was in the temporary office that I used in our office building in Boulder, which we were about to lose, when my brother Edwin called to tell me that my father had died. It was a sudden thing; he had broken his hip and was operated on, dying the next day. I had talked with him on the phone after the operation, and he sounded okay.

I sat in stunned silence for a while. Dan came in and asked what was wrong, and I told him. We just sat talking for a few minutes. Dan had met my father a couple of times and occasionally referred back to his conversations with the quiet mountain man. Barb, the accountant, came in, and when Dan told her, she broke into tears and rushed out.

I didn't cry then, but after a few more minutes I packed my briefcase and headed for the Denver Airport, where I was going anyway to meet JM, who was coming in from Houston for a planned weekend of fishing in the southern Colorado mountains. As I was riding along toward the airport, I was listening to country music, as usual, the Judds singing "Grandpa, Tell Me 'Bout the Good Old Days." That did it. I had to pull over to the side of the road for a while; I couldn't see to drive.

Speaking of crying, I saw my father cry only twice in my whole life. The first time was when the two older sons, Edwin and Kenneth, were first home from World War II. At either a Christmas or a Thanksgiving dinner, my father was asking the blessing when he broke down and cried as he thanked the Lord for bringing the boys back home. I have to admit, it startled me a little. I was only about eleven years old at the time. The second time was when his father died. Grampaw Hayes was pretty much of a n'er do well, but my father looked after him, building him a small house near ours on the same two acres where the old man lived out his final years. Grampaw died when I was at Berea College, and my father called to ask me to come home and sing at the funeral. When I got to the house, he was back in the bedroom by himself crying. The group at Grampaw Hayes's funeral was quite small, but that was one of the toughest singing assignments I ever had.

My father was a very generous person. In his later years, he shared the wealth of his sumptuous garden with anyone who needed it. Because he had a pure heart, he never shackled or chained me with the bonds of hate and prejudice. He never built any walls between me and the rest of humanity. I grew up totally unconstrained, and because of that, growing up free in his family in those mountains, I believe I was able to become more there than I could have anywhere else. When he was fourteen years old, he was working in a sawmill on the East Fork of the Pigeon River. When I was fourteen years old, I was playing football and baseball at Oakley High School.

He had his troubles, like so many others, in the Great Depression. However, he always provided for his children and owned his own home, which afforded us a good shelter to grow up in. When he died he left money in the bank. He was a tough guy; he never quit.

We had some difficulty getting home the weekend after he died, arriving a little late for the wake (they called it "the visitation") in Asheville on Sunday night. We were driving a beat-up blue Chevy S-10 pickup truck, the only vehicle left in our possession. Either nobody noticed or they were too kind to ask about where the Mercedes and the station wagons were. I never talked to my family about my financial condition, whether it was good or bad. Don't know why.

We buried my father in a grave on a hillside not much more than a mile from where I was born. My mother's brother, Luster, was there, talking loudly to anyone that would listen to him. Everybody looked at my Uncle Luster like he was crazy. They didn't understand that Luster loved Norman like he was a brother. The two old men had gardened the same bottom land below our house together for the past twenty years, as they talked about working in the mill and fishing and the mountains. They both slaved in the textile mill for decades, but their first allegiance was to the mountains and to the streams. They had passed that devotion along to me years earlier. They loved the mountains as if they were a part of them, which I believe they were then and still are now. Thank you, Spirit Father, for letting those two men touch my life.

Walking to Dunbar

Thursday, 21 February 1990—
Charleston, West Virginia

I was on a six-mile walk from downtown Charleston, West Virginia, to a suburb called Dunbar to visit a fly shop. On that long walk, I passed by a kaleidoscope of cultures plastered against the West Virginia hills. I walked by a housing project with mixed races and dilapidated cars with Georgia and New York license plates. Women in long black cars slowed down and waved to me as they passed. It was raining off and on, so I would occasionally have to duck into some doorway until the shower ended.

I was just passing the afternoon while waiting to give a talk on Hurricane Hugo before the Appalachian Geological Society that night. Hugo crossed the South Carolina coast near midnight on 29 September 1989, creating havoc in its wake, especially for the condos and other dwellings built too close to the beach. The extensive pine forests in the hurricane's path were demolished all the way to the North Carolina border. I made up a slide which said Hugo killed enough trees to yield lumber to rebuild all the houses in West Virginia.

The talk before the Appalachian Geological Society, as well as the one for the Pittsburgh Geological Society the night before, was for free, no money there. It was just an ego trip, kind of a favor for some of the students who had taken my American Association of Petroleum Geologists (AAPG) training seminar on the South Carolina coast.

During that long walk, I had a good chance to think things over. With the failure of the Canadian firm to buy out the energy side of the company, things were looking very bad indeed, and I thought the Boulder office was probably on the way out. It was obvious that we had to concentrate on building the environmental arm of the company to the point that it could be totally independent. That would take some doing.

Three weeks before that walk, in early February, I made my first trip to the *Exxon Valdez* spill site. JM and Jerry Galt had established eighteen permanent survey sites for NOAA in Prince William Sound for the purpose of monitoring the changes in the oiling over time, including its chemistry. However, another company had won the bid to do the monthly monitoring over the winter of 1989–1990. JM was there to overview their work for NOAA. I was there to study the penetration of the oil into the gravelly sediments. The oil was still present in considerable abundance in places, which meant that someone with my experience with gravel beaches might be of some assistance. I didn't get to see very many of the stations on that particular trip because we were snowed in at Valdez most of the time—a record thirty-four feet of snow by mid-January. It snowed three feet while we were there. However, the two intense days of field work that we did get in gave me lots to think about. Consequently, the science of gravel beaches was very much on my mind on that walk to Dunbar. We had already decided that I would join the inspection tour in April, and NOAA had plans for an all-out resurvey of the study sites in May, with RPI in charge of the geomorphological surveys from then until the end of the study. I was definitely going to be a member of that team. Therefore, despite all the bad business news, I had something new and exciting to think about.

When I got to the fly shop, I admired the three-hundred-dollar fly rods and other toys in the store, but all I bought was some fly tying materials with the money I "saved" by not spending cab fare. It rained on me all the way back from the fly shop to the hotel, and I was soaking wet when I got there.

The Great Rock Washer Debate

Saturday, 28 April 1990 —
The beach at Yalik Glacier, Kenai Peninsula, Alaska

I looked up from the trench I was digging through the asphalt pavement. Having emerged from their multicolored helicopter, five or six orange-clad figures were working their way toward us across the graveled surface of the upper beach. The chopper was mostly silver, with purple, yellow, and red stripes. Back in Homer the day before, our helicopter mechanic had referred to it as a "real pimpmobile."

"Who the hell is that?" I asked Russell, who had walked over to check out the trench.

"That's the Exxon big brass, according to John," was his reply.

I resolved to myself to put some distance between me and corporate America and walked down to the intertidal area where Daniel was examining the kelp beds. I knew John Dean, the Exxon representative on our team, would be tied up for a while before we could leave. Then I walked as far onto the submerged rock terrace as my waders would allow to check out the newly arrived Canada geese, goldeneyes, and Bonaparte's gulls with my binocs.

We were near the last of a magnificent ten days of working the outer Kenai Peninsula and Barren Islands as part of the inspection team checking out the *Exxon Valdez* oil that was left after the non-summer storms had reworked the shoreline. The purpose of the inspection, which involved over fifty people working on a number of teams, was to provide information for the planning of cleanup activities during the summer of 1990, the second summer after the spill. The weather was great the entire time, with seven or eight crystal-clear days.

The team leader was John Dean, another mountain man from West Virginia, who was a contract well-driller for Exxon in West Texas before this assignment. He was a pleasure to work with, and naturally, the two hillbillies, who were both in their mid-fifties, had a lot in common to talk about. Russell Kunibe was a biologist who represented the state of Alaska. Russell had been working the spill since day one and had an unbelievable nose for the oil, invariably finding it when none of the rest of us could. I represented NOAA and the U.S. Coast Guard on the team. I was also learning all I could about the subsurface oil and checking out places we could extend the NOAA surveys to later in the summer. Russell and I had contests to see who could dig the deepest trenches in the gravel. It was hard to excavate much deeper than a meter because of cave-ins.

It was an extremely enjoyable and historic time being back in Alaska. One afternoon we took a break by landing the helicopter on an ice field on top of the divide between Cook Inlet and the Pacific Ocean. Once out of the helicopter, we ran around waving our arms like kids trying to touch the sky in the exquisite blue air.

Occasionally, when we were in transit or back in the motel, I reflected on what was going on down in Colorado and South Carolina. You know that story by now. Nothing had changed for the better—yet. I told John, "I like this a lot more than what I've been doing for the past three years."

I stopped looking at the birds when I heard one of the young fast trackers from Exxon exclaim, "It sure is beautiful out here!"

"Right. Right." I thought, as I looked up at the snow on the mountain peaks on both sides of the ice-carved valley. "This is for sure a day of days on the Kenai Peninsula."

We were all gathered on the delta of the Yalik River, which is cutting through the moraine left behind as the Yalik Glacier retreats up a valley rarely visited by anything but brown bears, maybe some salmon, and other assorted wildlife. We were there to examine an asphalt pavement, gravelly sediment cemented in place by five or six inches of hardened oil, that was nearly a thousand yards long and ten or fifteen yards wide. The pavement was located on top of a relatively sheltered gravel bar that had formed on the delta surface after it sank several feet during the Good Friday earthquake in 1964. We were on the downthrown side of the fault line, the side that went down during the earthquake, unlike the other side in Prince William Sound, which went up. The oil cementing the gravel in place had been dumped on the delta surface in the form of a large patch of mousse a little over a year earlier, penetrating down into the gravel once the tide went out.

I noticed that Russell was digging on the far end of the pavement and went to join him. On the way, I passed by a German-born engineer, Hans Jahns, who was working on cleanup problem areas for Exxon. I met Hans earlier in a meeting in Anchorage, and we exchanged grunts as we passed each other. He was grimacing as he paced off the length of the pavement, no doubt thinking about the costs involved in removing this little piece of real estate to a landfill in Oregon.

As the Exxon executives were filing back into the helicopter, I walked over to John and asked, "We about ready to go?"

"Yep."

"The bosses, eh?"

"Yeah, and there's not a real person in the whole bunch." So said John Dean, the hillbilly well-digger out of Midland, Texas, on the afternoon of 28 April 1990.

John Dean told me a story about what happened when he was asked to manage some of the response activities for Exxon, which he agreed to do with alacrity. It sounded more interesting to him than drilling dry holes in West Texas.

"They picked me up in Midland in the corporate jet, and we were an arrogant bunch as we headed for Alaska to straighten everything out. But when we got here, and flew over all that oil going everywhere in the Sound, I knew we were in for more problems that we had ever imagined. In fact, I thought I was going to throw up."

I think it was on our way back to Homer that very night when, as we circled over Katchemak Bay, John pointed toward the shoreline. "See that rack of two-by-fours sticking up out of the water over there?" I looked in that direction and saw the remnant of what appeared to be the beginnings of a shed or something built out in the water a ways.

"What's that?"

"Well, it's really not anything at all, but I paid fifteen thousand dollars for it."

It was one of many schemes proposed to Exxon to help in the cleanup effort. John said it was more efficient to pay the cash than to deal with the politics of saying no. Many schemes and dreams were proposed, but I wonder if any were as grandiose as the "rock washer."

I cringed every time I heard the term *rock washer*, because it was a scientific misnomer. This "rock washer" was a scheme proposed by somebody in the state's employ to clean up the oiled *gravel* beaches, not the rocky shorelines. Okay, so who cares what it was called? I care, I'm a sedimentologist, remember?

The plan was to locate a giant wash basin offshore in a barge connected to a conveyor belt that would carry the oiled gravel, some fragments of which Olympic weight lifters could not throw ten feet, to the wash basin where it would be cleaned. The cleanly washed gravel would later be returned to the beach unharmed. Sound preposterous? Well, I believe that is what the Exxon team thought. I know that is what some of the NOAA managers thought, but nobody could think of a cheap way to say no, at least not that the state of Alaska would buy off on. Therefore, the "NEBA from hell" was born.

NEBA stands for *net environmental benefit analysis*. It was supposed to be a routine that a committee goes through to determine if the environmental benefits of some activity—in this case, the rock washer—justify the costs. As far as the rock washer itself was concerned, the state of Alaska wanted it, Exxon didn't want it, believe it or not, and our client, NOAA, was caught in the middle, as usual. So NOAA was chosen to chair this NEBA committee, and guess who was assigned that task? None other than JM.

NOAA asked me to attend a NEBA meeting right after the field survey just discussed. That meeting, held on 1 May 1990, was part of a continuing series of meetings on the subject. JM had been in Alaska the same time as I had, participating on a different shoreline inspection team aboard a boat in Prince William Sound. The NEBA meeting was scheduled to occur when the surveys ended, which was during the neap tides. Neap tides occur during the first and third quarter of the moon and are the smallest tides of the month. I was there to chair a work group of sedimentologists, who were supposed to decide if the rock washer would harm the sedimentological integrity of the beach, among other things.

The meeting took place in a conference room in the Exxon offices in Anchorage. Despite the fact that the representatives of the Alaska Department of Environmental Conservation (ADEC) were late for the meeting, JM was just about ready to start it anyway. Introductions were made all around. The Exxon geological engineer sitting next to me reminded me that we went to graduate school together at UT. He was a couple of years behind me; his name was Ted Bence. He looked about the same, except that his hair was quite gray. What? No Grecian Formula? I have to admit that I was using it then, but I quit after the airplane crash, which was coming up in three months. That was one of the things I promised the Lord I would give up when I was crawling up that rock. Only kidding!

Just as JM was starting to go over the agenda for the meeting, the ADEC mafia stormed in, led by one short, red-headed ex-New Yorker I will call "Newk." The first thing Newk did was walk up to JM, slam a file down on the table in front of her, and say in a loud voice, "Did you really say what was quoted in the paper?"

The file contained several scientific articles on sediment washing at oil spills. JM had been quoted in the paper saying rock washers had not been used at oil spills before.

"I said in the U.S.A.," was her quick response.

Then there was a break, while the new folks, including me, were introduced to the late arrivals. I really wanted to punch Newk out, not shake his hand, but good sense prevailed. Or did it?

John the Exxon engineer, who had just flipped through the articles in the file Newk had thrown on the table, said to JM, "These papers are all on sand washing." I, for one, was glad to know that John knew the difference between sand and gravel, but that bit of factual information didn't seem to phase the red-haired Newk, who continued to loudly badger the Exxon representatives throughout the meeting. I bit my tongue to keep from telling him to shut up. I probably would have if JM had not been chairing the meeting. I did, however, at one point request that Newk not be assigned to my work group. Newk was reassigned.

Exxon was resigned, having already agreed to spend over a million bucks on a rock washer model. The meeting ended, and we left. "All of the meetings are like that," JM told me as we walked away.

Needless to say, at that point in time I was leaning in favor of Exxon. Newk made me mad, but as time passed, I became more sympathetic to his cause, in an emotional sense. Why? Because the state was overmatched. It was the ragamuffins versus the Bronx Bombers. Ed Owens at one time had over forty geologists working on his team for Exxon. I think the state had two or three at best. And the salary structure . . . well, you know.

I thought of a great analogy for this contest. Berea College did not have fraternities, nor was there that much emphasis on athletics. The school participated in only four sports at the intercollegiate level, one of which was baseball. Intramural sports, though, were contested heatedly. The nearest thing to fraternities were the efforts by individuals on certain floors in the men's dormitories, the basis for the intramural teams, to recruit the best athletes, based on seniority. That is, on the floors that wanted to excel in sports, the seniors would list one of the best freshmen athletes as their roommate each year, switching back to their original roommate as soon the fall semester started. I was recruited by Howard A (the basement) one year, not because of my skills at touch football, which Monarchy Wyatt didn't want me to play because of my trick knee, but because I was a sure winner in the horseshoes contest each spring, which were points toward the intramural championship just like in basketball or any other sport. My father, the horseshoe champion of the Enka textile mill, taught me how to pitch horseshoes. Howard A was always in the top two in the campus intramural standings, whereas Howard Two, which was where all the upper-class intellectuals gravitated to, was always on the bottom. One time Howard A played Howard Two in a basketball game, and as the spectators were walking back to the dorm afterward, I asked one of the physics majors from Howard Two the score of the game.

"Several to a few," was his reply. The score was something like seventy-five to eighteen, in favor of Howard A, of course.

Exxon was Howard A and ADEC was Howard Two.

We had many discussions, meetings, and communications about the rock washer, eventually writing a report. Exxon called in its paid consultants, staffed by Ph.D. chemists with resumes as long as the Chicago phone book. These consultants took it upon themselves to spend long hours going over our report word by word, affecting any changes they could squeeze out in Exxon's favor. Newk's antics were nothing compared to this, which was unacceptable!

John the Exxon engineer kept referring to the rock washer process as "ripping the hell out of the beach." I guess this "assessment" was supposed to mean that washing the gravel was a bad thing to do to the beach. The geomorphologists and sedimentologists on my work group laughed at this assertion. Mother Nature "ripped the hell out of the beach" several times a year during a process called "coastal storms."

One of the items that the chemists hired by Exxon continually changed in our report was our contention that the oil in the subsurface sediments on the beach would continue to sheen over time.[36] This meant that some oil, however small the quantities, would still be available for organism uptake if the oil were left in the subsurface gravel. They insisted that the oil would not sheen after one year.

From the report JM and I wrote for NOAA based on our summer of 1994 survey at station N-3 on Smith Island, which was carried out five years and four months after the spill: "The subsurface oil generated chronic sheening at this site, visible as silver sheens in the ground water draining from the beach during the falling tide." And we had color photographs and chemical samples to prove it. The NOAA biologists found that blue mussels located near that beach were bioaccumulating traces of oil for several years after the spill.

The rock washer was never built. In all fairness to ADEC, however, we could find no *bona fide* scientific reasons in our studies for disallowing it as a cleanup technique.

Howard A = 75; Howard Two = 18.

In the summers of 1990 and 1991, Exxon sponsored a series of berm relocation projects during which they "ripped the hell out of" about three thousand yards of beach. In that case, they just pushed the oiled gravel down the beach a ways, and let the storm waves clean it and push it back up where it came from. This process was a lot cheaper than the rock washer. If we had done a NEBA on berm relocation, I am sure we would have found no *bona fide* scientific reasons for disallowing that cleanup technique either.

During May and June of 1990, JM and I ran an extensive resurvey of the NOAA sites and established a few new ones on the Kenai Peninsula and the Barren Islands. I don't want to imply that going to oil spills is ever fun, but as far as the science was concerned, this was a wonderful trip for me. We invented new ways of describing gravel beaches and have published several papers on the results.

I was scheduled to go on the September 1990 field survey, but the plane crash changed that. JM went and was able to see firsthand the results of some of the berm relocation projects, which, for the most part, were quite successful. The research on the tricky subsurface oil continues even to this day. Many of the NOAA geomorphology/chemistry sites have been resurveyed fourteen times. During a survey carried out in July 1997, we found still only moderately weathered oil in the subsurface sediments at several sites. Therefore, we are still learning from this spill.

The NOAA biologists, led by Alan Mearns and Gary Shigenaka, worked some of the same stations as ours, as well as several others, carrying out much more extensive studies than the ones we did. One of their main conclusions was that the hot-water washing process that was executed on a grand scale during the main cleanup effort in the summer of 1989 slowed down the recovery rates of the intertidal biota. Exxon's "hired guns," with resumes as long as the Chicago phone book, did not agree with that conclusion, either.

> Brother, do not believe that I came here to get presents from you. If you offer us any, we will not take them. By our taking goods from you, you will hereafter say that with them you purchased another piece of land from us.

Tecumseh

. . . another piece of our minds from us.

Me

The Native Peoples and the Spill

A closed mind is a wonderful thing to lose.

Bumper sticker seen in Columbia, South
Carolina

Sunday, 4 March 1990—
Anchorage Airport

JM and I spent the last few days at a conference in Cordova on proposed research for the future in Prince William Sound, particularly with
reference to the *Exxon Valdez* oil spill. I thought the meeting was great.
There was a bit of whining about the lawyer's muzzling some of the
speakers, but it was a side issue. One of the things that struck home to
me was the effect of the spill upon the native peoples of the region, who
were in conspicuous attendance at the meeting. I wrote in my diary,
"This is something I never would have appreciated without being here."

John Robinson of NOAA, the veteran of many spills, told JM about his
eyes being opened to the impact of the *Exxon Valdez* spill on the native
peoples when the following incident happened. During the first summer
after the spill, John was touring the native villages in the Kodiak Island
area on a fact-finding mission. They landed their float plane near the
shore at one of the villages. John asked a young boy if he would wade out
and pull the plane closer with a short rope they would toss him. The boy
wouldn't go near the water because it was not safe. The sacred, life-
giving waters had been poisoned by the toxic spill. That boy was not the
only one of the native peoples who felt that way.

Not too long ago, Bill Baird, a coastal engineer friend of mine with
whom I collaborate occasionally on beach erosion projects, asked me if

I had ever read the book *Ishmael* by Daniel Quinn.[37] He also recommended the book to everyone in his own engineering firm. Bill's motto for their engineering activities is to "make the world a better place to be."

In *Ishmael* and its sequel, *The Story of B,* Quinn makes the analogy between the way we have destroyed and continue to destroy the native cultures with the way we are destroying the natural environment. He calls the native peoples the "Leavers" and us the "Takers." That concept may be too simplistic for some, but it's good enough for me.

The Mother of All Spills

The chronology of events surrounding the Gulf War, and its associated largest oil spill in history, is as follows:

2 August 1990	Iraq invades Kuwait
16 January 1991	Allied forces declare war on Iraq and begin bombing Baghdad
19–23 January 1991	Iraq releases oil from Sea Island Terminal (in Kuwait) into the Gulf
9 February 1991	Smoke plumes from oil-well fires visible on satellite imagery
23 February 1991	Allied ground offensive begins
3 March 1991	Conclusion of ground campaign

When the Iraqis invaded Kuwait, two good friends of mine were trapped there. One was Mohammad Al-Sarawi, my graduate student, who was then a professor in the geology department at the University of Kuwait. The other was Paul Pawlowski, the chief architect on the Kuwait waterfront project, who I had worked closely with for several years in the late 1970s and early 1980s.

On the night of 8 August, six days after the Iraqi invasion, Mohammad loaded his wife and two small children into his jeep and followed an Iranian underground operative through the desert to the Saudi border. Being a geologist who worked often in the field, Mohammad naturally owned a four-wheel drive vehicle, unlike other would-be escapees who tried to drive through the desert dune sand in their Mercedes and Buicks.

"Dozens of those cars were buried in the sand up to their windows, Miles," is the way Mohammad described the spectacle he encountered there in the desert in the dark.

Meanwhile, Paul's Kuwaiti friends hid him in their basement for several weeks until he also was led to safety by an underground operative. I asked Paul what he did there alone in that basement for all that time. "Played with myself," was his curt reply.

On 12 August, while I was crawling up the rock to a safe spot where I would wait for Jamie to save me after the airplane crash, Paul was "playing with himself" in the basement of a private home somewhere in Kuwait City. Four days earlier, Mohammad was driving through the dunes in the dark to safety across the Saudi border. I guess you could say that August 1990 was a pretty eventful month for all three of us.

Wednesday, 16 January 1991—
Columbia, South Carolina

Despite the fact that I had been walking only a short time since the plane crash, JM and I decided that I should join the NOAA field survey of the *Exxon Valdez* stations in Prince William Sound scheduled to take place between 18–26 January 1991. She arranged for us to conduct the survey aboard the *U.S.S. Sweetbriar*, a U.S. Coast Guard buoy tender which was scheduled to carry out a routine mission in the Sound at that time.

On Wednesday morning, 16 January, we left Columbia headed for Anchorage. After our plane arrived in Seattle, we walked into the terminal, which was quiet and subdued, unlike its usual hustle and bustle. We heard someone whisper that U.S. planes were bombing Baghdad. Then we stopped outside a bar, along with dozens of other quiet folks, straining to hear Dan Rather explain that they really didn't know what was going on over there in the dark.

As we walked to catch our plane to Anchorage, I felt terrible, heartsick. "Why are we doing this?" I thought.

Saturday, 26 January 1991—
Onboard the U.S.S. Sweetbrier *in Cordova Harbor*

"Reveille! Reveille! Reveille! Up. Up. Up. Up. Haul out and trice up! Now. Up. Up. Oh, six-thirty. Reveille!"

That was the call I heard every morning on the ship, except this one, which was a day off for the crew. I was up at 6:30 A.M. anyway, packing my duffel for the trip to the Cordova Airport to meet the Exxon Funding

Committee plane for a quick ride to Anchorage. We were scheduled to make a presentation of our findings to the *Exxon Valdez* Technical Committee at the Exxon offices later on that morning. At the airport in Cordova, we packed the plane in the dark, and seven of us got in, taking up every available seat.

"How was the flight?" the Exxon security guard who met us at the Anchorage Airport asked.

"Only one hour and twenty minutes of sheer terror," I replied.

Except for the rescue helicopter, this was the smallest plane I had ridden in since the crash. It didn't help any that the altimeter in the passenger area, which I watched apprehensively on our approach through a fog bank into Anchorage, registered −300 feet before they lowered the wheels for the landing.

As we rode toward the Exxon HQ, I reflected on the survey. No doubt, the *Sweetbriar* had been a great platform to work from. When the wind blew from one direction, the CO would simply order the ship to the opposite side of an island where we could do a lee station in calm water. We had our moments of excitement, nonetheless. One night the wind topped eighty miles per hour, and then there was the night we almost ran into the side of a mountain while making a practice run through one of the narrow channels. The weather was highly variable. One day we had sunshine, sleet, heavy snow, and a hard rain while we were surveying a single station.

We were transported to the beach stations by coasties driving inflatable boats. We wore rubber survival suits, called "Fitzrights," which I had a tough time getting into with my injured legs. Steve Lehmann, a NOAA scientific support coordinator from the New England district, got the job of assisting me. He referred to himself as "Miles's man." JM had fun in the Fitzrights, jumping into the frigid water and swimming along beside the inflatable on occasion.

We did twenty-five stations, not leaving out any of the planned work. The snow was waist deep behind the last high-tide line at some of the profiles, and the upper beach was frozen solid. But we found lots of oil in the subsurface gravel at several stations, which had changed very little in the past year.

Some mornings, I dreaded the pain of climbing into the survival suit and going into the field. I was unsure of my ability to perform physically. At times I was quite cold in my Fitzright, but I was able to get around better than I had any right to expect. The boat drivers worried me some, especially when the wind was up and they bounced the small craft through the big waves, but I eventually got over those concerns and thoroughly enjoyed the last few days. The last day, in particular, was a fine work day. Looking from the high deck of the ship in every direction, you could see the snow-capped peaks all around the Sound. I was sad to wrap up the work and leave one of the most beautiful spots in the whole world.

Several people from the Alaskan state agencies showed up for our presentation at the Exxon HQ. John the engineer and other associates from Exxon attended. JM had already written a short report on our results, which was distributed to the attendees. There wasn't much bite in the questions, and the meeting went well.

At the airport, we bought the *Anchorage Times,* which had the headline: "Iraq Unleashes Oil Spill!" It was rumored that this spill was much larger than the *Exxon Valdez.* It turned out to be twenty times as big. We learned that NOAA HAZMAT was headed to D.C. to deal with the spill. Could RPI be far behind?

JM and I stopped in Denver that Saturday night and rented a *free* Cadillac which we drove to Boulder, where the office was still barely functioning. We had an abbreviated management meeting Sunday morning because NOAA had summoned JM to D.C. for an immediate response to Saddam's giant spill in the Arabian Gulf. We had already been working on a proposal on a strategy for habitat cleanup response, restoration, etc., but we didn't know at that time what our role would be. I stayed in Boulder so Dan and I could have more fun with the Colorado banks and their legal counsel.

JM was in D.C., and back in Columbia, the three local TV stations were conducting interviews with RPI's Linos Cotsapas and Jeff Dahlin on our work with oil spills. We were not supposed to talk about the details of our work on the Gulf spill, which was just beginning. JM and I were in the process of obtaining SECRET clearances so we could participate in the response activities.

"Can we come back when you can talk about the spill in the Persian Gulf?" the TV interviewers asked Linos and Jeff.

"Hurry home and get these people off my back," Jack Moore, our business manager, pleaded over the phone, and I headed back.

Next day, Wednesday, I went through the big pile of mail on my desk and found a letter from Mohammad Al-Sarawi who, as noted earlier, had escaped from Kuwait six days after the Iraqi attack and was in England with his wife and two children. He said, "Best regards and love to my adviser who helped me a lot, a lot, a lot that I will never forget."

Up at 7:00 A.M., I headed to the British Petroleum station for a breakfast of Moon Pies and Diet Coke. What? Don't worry, I just eat like that when JM is out of town. I had a 9:00 A.M. appointment with a local radio station to talk about oil spills. I had agreed to this appointment when I was in Boulder. At that time, I didn't know it was a call-in talk show, thinking they would just tape an interview.

The station was on a dead-end road in the vicinity of the Lower Saluda River, and I arrived a little early. A little ways back down the road I spotted a fishing and portage spot on the river I had never seen before, so I went back for a closer inspection. I was thinking about going fishing on Saturday, which would have been my first fishing trip since the crash. However, the river was high and slightly discolored, almost out of its banks—not too good for fly fishing.

I was shivering when I returned to the station because I was wearing only a sweater. The wind was blowing across the water, and the temperature was below freezing.

"So, you just got back from Alaska?" the young lady who had invited me asked. I felt kind of silly, shivering so badly. I was, after all, the Alaskan explorer returned from conquering the frigid wilderness. She told me the program was a talk show (surprise!) that would last for an hour, and that John Risley would be the host, and by the way, you can sit right here in any of these chairs. All of these microphones work.

"The show starts at 9:06," she said as she handed me a cup of coffee, which I didn't drink. The Diet Coke, the wind, and the shock of being on a talk show for the first time in my life had already awakened me sufficiently, thank you. I sat alone in the studio for a couple of minutes, wondering what the hell was going to happen next.

"Hello there, young man!" John said as he breezed in. Well, I guess in comparison to old John, I was still pretty young.

"It's better if we don't discuss this too much before we start. It will seem more natural that way," he said as he sat down on the other side of the console and adjusted his head set. Good thing. We only had about thirty seconds left until show time.

First, he asked me about the Alaska trip. I told him about the day we had clear sky, a blizzard of huge snowflakes, sleet, and a downpour of rain all in the span of a couple of hours while we surveyed one of the stations. We also discussed the spill a little bit.

Eventually, there was a brief CNN news report on the war in the Gulf. I mentioned my airplane crash as we chatted while the news was on.

Then the phone calls started. A caller, Bill, was upset that the wildlife service was shooting birds and throwing them in the water to mimic the drift of the dead birds killed by the *Exxon Valdez* spill. I assured him that I had never heard of this "costly" project, but if it had actually happened, it was probably instigated by lawyers. After my stay in Boulder, I was not particularly high on lawyers.

During the slack periods between calls, John asked me about the crash. Thus, a few hundred more people heard that story. I was impressed by the way John deftly switched topics, introduced the callers, and switched to the two news briefings, all as smooth as silk.

"You ever thought about writing a book on this stuff?" he asked right out of the blue.

"Uh. Well, yes. I have." I stammered, hoping we would quickly get another call.

Another Bill asked about cleaning single "rocks" with paper napkins. He meant cleaning fragments of gravel, but I didn't think the audience would perceive the difference as a significant one so I didn't correct him. What Bill described seemed to echo a lot of people's image of how Exxon cleaned up the *Exxon Valdez* spill.

"Well, actually, once they got started, they did a very good job," I said. And then I tried to explain Exxon's omnibooms and maxibarges. Good luck!

"And this man does not work for Exxon!" said John.

Then another caller asked about the new species of maxi oil eaters that she had heard about on the news. She was referring to some kind of super bacteria that a group of hustlers had been talking about dumping on the oil that was presently floating around in the Gulf. I had read about their scam in *USA Today*.

"Well, I've been thinking about that," I replied. "If those things really can eat up that much oil, I don't want to be on that side of the world when they get through with the oil."

"Yeah, I was thinking something like that myself," the caller responded.

"It won't work," I muttered halfway under my breath, being partially drowned out by the "click" she made as she hung up the phone.

Then a gentleman I presumed to be an Exxon shareholder called. He was angry about Exxon spending over three billion dollars on the cleanup. He had been doing some figuring on the back of an envelope.

"Does this mean that if the spill in the Arabian Gulf is forty times as large as the Alaskan spill (it was actually twenty times—this guy's exaggeration factor was as big as mine!) that they will spend $128 billion to clean it up?"

"Well probably not, they do have coastal roads in Saudi Arabia and Abu Dhabi," I said. I had some trouble with that one.

So it went. We eventually ended the hour by talking about what we are going to do for an energy source when we run completely out of oil about fifty years from now, not that old John and I had much to worry about on that score. And, of course, I was asked to render my expert opinion on how the Saudis would get water if the desalinization plants were oiled. I didn't know it at the time, but some of my future Saudi students and co-workers were busily placing multiple barrier booms around the intakes to the world's largest desalinization plant at Jubail, Saudi Arabia, at that very moment.

Everybody back at the RPI office had listened to the broadcast. They all smiled as I walked to my office. Maybe they liked my jokes.

Later, during one of our multitudinous conference calls, I told Dan and Jack, "Well, this is weird, but it beats the hell out of talking to creditors." They wanted to switch places.

The next week I joined JM in D.C. to help with the response to the spill, using our new SECRET clearances to view the burning tankers and other aspects of the spill on some of the latest satellite imagery. The military was considerably interested in our ESI maps of the Kuwait shoreline, which included data giving beach slopes and trafficability, among other things, around the Kuwait City waterfront. We assumed they were using that data to plan a beach landing for the offensive.

All the old-time NOAA responders were there. John Robinson, who had recently moved to D.C., was in charge of the response to this, the "mother of all oil spills."

One day, John took Jerry Galt and me to a CIA shop obscured by storefronts on a back street. A large number of people were busily working on all aspects of the Gulf War and its associated spill. We two old-timers were supposed to brief the assigned CIA staff on the technical aspects of oil spills. They had a lot to learn, and Jerry and I, never known to hide our lamps under a bushel, gave them all the hot poop we could think of in the short time allowed.

That night, John and his friend Francesca Cava gave a party at their place. JM and I regaled Eric Schneider, a geologist I had known for a long time who was a Washington veteran, with the story of the crash. "I've never heard a story like that in my life!" he exclaimed.

Then we told whoever would listen stories about our home on Deep River Bluff in the backwoods of Calhoun County, South Carolina. We talked about rooster farmers, the sad fate of the hunting dog we had befriended called "No. 6," and drunken hunters lost in our swamp. After we finished our stories, John said, "It's time to do another 'Suzanne's Lament!'"

It was odd to be back in that loop again, after chasing the energy business for so long. We laughed and laughed when JM told the story about my response to the concerns expressed by the bureaucrat about "greased beach berms sliding into the ocean" at the Christmas 1978 spill in Puerto Rico. A little later, John said, "Hey, you guys missed the Miles and Jerry show at the CIA today."

Midging the San Juan

We were on a Delta jet headed for Denver. As we took off, the guy next to me said, "Look at all those Eastern planes, amazing." One of the terminals was completely surrounded by idle Eastern jets.

I responded, "Yeah, they just walked away and left the lights on, didn't they!" I learned that phrase from our lawyer, Billy Robinson, in one of his lectures on the different techniques for declaring bankruptcy.

As we approached the Denver Airport, I wrote in my diary, "Flying in these planes still terrorizes the living hell out of me! Especially turbulence, which is what we are now going through as we go down. It doesn't bother me . . . bother me . . . bother me . . . bother me . . . bother me . . . bother me!"

I was on my way to Denver to meet Dan for a couple of days of fly fishing on the San Juan River below the Navajo Dam in New Mexico. It had been nearly seven months since I had been fishing, and I was dying to try out the new nine-foot, four-weight Sage fly rod Dan and the gang had given me as a get-well present after the crash. I don't particularly like fishing the San Juan because it is always so crowded by people trying to catch the huge trout in the even-temperatured waters in the tailrace below the dam. The only other choice we had in late February was the South Platte River below a dam south of Denver, but let's face it, the fish there are too educated for me to catch. Also, I love New Mexico, and this trip gave us another chance to drop in on an old acquaintance, Miles Sanchez, who owned a motel in Chama, up near the Colorado border. Not to mention a visit to Rita's Mexican Restaurant, which was located right across the street from Miles's place.

Dan met me at the Denver Airport, and we headed south in a small white Ford Escort he had borrowed from his father. My legs would still get stiff sitting in the cramped quarters of the small car. Sometimes I had to lie down in the back and stretch my legs over the top of the front seat.

Of course, the war in the Gulf was foremost in our minds. The Desert

Storm offensive was under way, and I wondered if they really were going to use our ESI maps to plan a landing along the Kuwait City waterfront. They didn't, of course, choosing to circle around the city above a high escarpment and over into Iraq the back way. Like everybody else, Dan and I became a little misty eyed each time we heard Lee Greenwood's rendition of "Proud to be an American" on the radio.

In Chama, we found that Miles had flown the coop, but his wife was glad to see us, and the Mexican food was still the same high quality we had fantasized about. That night, we watched the goings on in the Gulf on CNN, the brainchild of the one and only Ted Turner.

Next morning, we were fishing the San Juan by 9:45 A.M., along with what appeared to be about a hundred other fly fishermen. I counted twenty-five cars parked at the Texas Hole, where we fished, and ten at the parking area for the Lower Flats. The temperature was in the forties, and it was a little windy, but the sky was crystal clear. It was exhilarating to be fishing again. Thankfully, the wading was easy in the fine gravel of the shallow braided channels.

I wasn't fishing really great; my casting was terrible, and I kept having trouble with my leader knots. Although we fished until 5:00 P.M., we caught only two fish that day, but they were both whoppers, the only things I saw in that river.

About 10:00 A.M., we watched hordes of the big fish come out of the major channels into the shallow flats to feed on what we surmised to be No. 26–28 black-bodied, gray-winged midges. First the fish were feeding on the midge pupae just under the water surface. Later, they started feeding on spent adults with the wings flat on the surface, some of which had trailing pupal cases. We knew what they were feeding on, but we had no flies to match the hatch, and those fish were very selective. They had learned by bitter experience not to hit the clumsily tied larger versions of the normal dry flies that could easily fool less-educated trout in other streams.

We spent the night in a motel in Farmington, New Mexico, while Desert Storm raged on. With one eye on CNN, I tied several variations of the midges we had seen on some No. 24 dry fly hooks that I had in my fly-tying kit. I tied a version with a flattened wing, one with a trailing shuck, and a standard Griffith's gnat which had a hackle wrapped all around the hook. We called the new flies "Mo-Dan midges" (except for the gnat). For the record, a No. 24 hook is approximately one-quarter of an inch long. Many of the huge trout we had seen feeding on the tiny insects I was trying to mimic were well over twenty inches long.

Next morning, we chatted for a while with a farmer who had a dummy of Saddam on one of his tractors parked by the side of the road. A large American flag was stuck up Saddam's backside. The farmer joined us as we admired his art work, and we discussed the progress of the war. It sounded like a rout so far.

We were fishing near the Texas Hole by 8:45 A.M., but it was 9:30 before the big fish came onto the flats looking for the midges again. At that time, I was fishing with a Mo-Dan midge with the flattened wings but was getting no strikes. I could see that the fish were feeding under the surface on pupae, none of which I had tied the night before. Then Dan and I walked back to the car where I took my scissors out of the tying kit and trimmed the wings off several of the flat-winged flies, creating a midge pupae imitation. Back on the water, I greased the leader with fly floatant, leaving the last nine inches above the fly ungreased, which allowed the fly to hang vertical in the water just like the natural pupae. I caught three fish with this rig over the next hour, one of which was twenty inches long. These waters are catch-and-release, but I would have released them in any event.

About 10:50 A.M., the fish started feeding heavily on the surface, so we switched back to the dry flat wing and caught four more. Around noon, when the hatch was so heavy that our low-floating flat wings were lost in the crowd, as it were, we switched to the Griffith's gnat, which was more visible to the fish and caught one more.

We quit at 12:30 P.M., because it looked like the weather was going to change, and we had to go through a high pass on our drive back to Boulder that night. In total, we had brought to hand eight of the big fish in a little over two hours of fishing; two of the fish were over twenty inches long. The fact that we had figured out the hatch and were able to match it with the small Mo-Dan midges made it one of our more satisfying fishing days.

Driving up toward the mountains, I played the tape I had recorded during the trip. The recording included numerous erudite sayings, such as "It was remarkable how well those No. 24 hooks held those big fish!" and "I was amazed how big the fish were, yet feeding on those tiny midges . . . The trailing husks were a joke . . . that was a bad idea." And so on.

We were listening to the radio as we approached Walsenburg and the interstate, and we heard President Bush announce that Desert Storm was over. They were pulling back and not going all the way after Saddam! Over another Mexican dinner in Walsenburg, we puzzled over why they hadn't finished the job.

"Probably the Saudis' idea," I speculated.

From Denver I went up to Seattle for a NOAA meeting on the results of the *Exxon Valdez* spill site survey, and then JM and I went to San Diego to attend the International Oil Spill Conference, where we gave papers on the *Exxon Valdez* work, among other things. As a bonus, we were able to visit with my daughter Joy and her boyfriend Steve, both of whom were star Ph.D. candidates at UC San Diego.

At the AAPG Conference in Dallas a few weeks later, I was walking through the exhibit hall when I spotted an old associate, Chris Kendall, standing outside his exhibit booth. When he saw me, Chris jerked himself erect, spread out his arms, and said in a loud voice, "Miles, you are supposed to be dead!"

He was not the only one who was surprised to see me walking around on my own two legs so soon after the crash. One of my geology heroes, Dr. Bob Ginsburg of the University of Miami, asked me, "What are you going to do with your second life?" Continue my quest to survey the coastline of the whole world, I guess.

I'm not going to write about it in detail here, but in April and October of that year, 1991, I participated in a beach erosion study of the entire coast of Oman. I spent many days traveling from one end of the country to the other, surveying its shoreline on the Indian Ocean with Bill Baird, Steve Sturm, and others. With the exception of Alaska's, that coastline, which ranges from thousand-foot-high cliffs, to coral reefs, to some of the most spectacular beaches in the whole world, was the most exciting I had ever studied. We slept on the desert shoreline at night looking up into the clear sky, feeling we could just reach up and touch the Milky Way and the Southern Cross. I've never felt more a part of our home galaxy than I did there on those Oman beaches in the cool night air. It all seemed like a reward from the Spirit Father for having survived the business and airplane crashes of the past two years.

Ramadan

In the beginning God created the heavens and the earth . . .
And God saw all that he had made, and behold, it was very good.

Genesis 1 : 1, 31

Wednesday, 11 March 1997—
On a sabhka in Dawhat al Mussalimiyah, Saudi Arabia

We came down over a rut pretty hard, hitting a low place with the front tires and sending a flood of water over the entire body of the Pajero. We drove on for a few seconds in a blackout as the water drained off the top of our vehicle. The water level over the track almost covered the tires. Even after the bulk of the splashed water had drained off, the racing windshield wipers were just barely clearing the downpour of rain. Ahead of us lay what appeared to be a lake of water covering the track, and in the distance it went on and on over the flat surface of the sabhka, until the horizon was obscured by the heavy rain.

"Oh excrement!" Abdul Halim exclaimed, the first time I had ever heard him say that, as another flood of water blinded him again.

Up to then, I had not been overly concerned, because Halim told me that he had never been stuck in a sabhka, only in the dunes. He slowed a bit as we rolled up out of the water onto a small mound. Then he looked over at me in the right front seat and said, "I have never drive on a sabhka this bad."

A sabhka is an extremely flat surface, usually a salt flat that is occasionally flooded by wind-generated tides, and it rims the shoreline where the hinterland is flat. The sediment is usually muddy sand and typically quite soft, making driving on it very tricky indeed. In the area we were in, the Saudi coast guard repeatedly drives over certain paths, driving out the ground water, until a hardened track is formed. If you venture too far off the track, down the tires sink into the soft sediment. It takes a real expert to drive on the sabhka for any length of time without getting stuck.

Then Halim stopped and signaled to the Saudi Aramco vehicle that was trailing behind us to stay where they were until we made an attempt to get through the lake in front. As we bounced and splashed and skidded our way through, I glanced in the back seat at JM and Ahmed, who appeared to be enjoying the ride.

Once we came up onto some harder and slightly higher ground, Halim turned the Pajero around and flashed the headlights at the Aramco people, indicating that they should come on through. I jumped out and took a couple of photos of them splashing their way up to us. Then we headed around the edge of the sabhka on the way to our station 19SA.

We were in Saudi Arabia for our third survey of some study sites we had established across the intertidal zone in the Jubail area, which was very heavily oiled during the Gulf War spill of 1991, six years earlier.

The Gulf War oil spill was by far the largest oil spill in history (216 million gallons), estimated to be twenty times as large as the *Exxon Valdez* spill. Much of the oil came from releases caused by the Iraqis at the shore terminals in Kuwait, but some apparently came from tankers sunk during the war. When the oil approached the Saudi Arabian shore, a major effort was made to remove it from the water surface, with four times as much as was spilled by the *Exxon Valdez* being collected and stored in large trenches in the desert. After that, however, little effort was made to clean the oiled shoreline. Instead, the oil was left for the vagaries of nature to do the job. Consequently, considerable could be learned from a detailed study of the different oiled habitats along the Saudi coastline with regard to the efficiency of natural processes in cleaning up such a large spill. This was one of the principal goals of the project we eventually carried out.

Once again, John Robinson had been instrumental in getting us involved in a major oil-spill study. After a long series of meetings and negotiations, NOAA had decided to furnish its research vessel, the *Mt. Mitchell,* to be used by the scientists of the Gulf countries for a major research cruise to determine what had been the effect of the Gulf War oil spill on the Gulf. The cruise started on 26 February and ended on 1 June in 1992. John, who was the technical manager of the cruise, had recommended that RPI take the leadership role for Leg II, which focused on the heavily oiled area around Jubail, an industrial complex located about halfway down the shoreline from Kuwait to the United Arab Emirates border. JM was the chief scientist for that portion of Leg II involved with the offshore studies. I was the chief scientist of the dunes, and I supervised the onshore research. Leg II was scheduled to take place between 16 March and 6 April. We wrote a proposal which was generously funded by the Marine Spill Response Corporation (MSRC), with a Ph.D. biologist, Don Aurand, serving as the contract monitor of the project for MSRC.

On the Friday night before that splashy ride across the sabhka, which took place on Wednesday, 11 March 1997, JM and I arrived in Dhahran, Saudi Arabia, about midnight, after a twenty-six-hour trip from Columbia. We were happy to see a white-robed Saudi wave to us from behind the customs gate. It was Khalid, one of the MEPA (Saudi Arabia's equivalent of our EPA) scientists who had spent six weeks at RPI for oil-spill response training a couple of years earlier.

Finally, we made it through customs to the outside, where we were met by Aziz al Omari, the boss of the local MEPA operation and also one of the students in the six-week training at RPI, Abdul Halim al Momen, another MEPA technician, the driver who drove us across the sabhka in the previously mentioned rainstorm, and Ron Williams, a physical oceanographer at Saudi Aramco, which was providing us with logistical support on this short visit. Though tired from the trip, we were glad to have the opportunity to visit the Gulf War spill site once again. We had studied the oiled shoreline one year after the spill in 1992 and two years after in 1993. We were anxious to see if our study sites were recovering at the rates indicated in the lectures I had listened to at the conference in Kuwait back in November (discussed in the chapter "A Primer on Oil Spills," earlier in this book). As we talked, I was thinking about how it all had started for me at a meeting held right there in Dhahran in December 1991.

Wednesday, 11 December 1991—
Domestic air terminal, Dhahran, Saudi Arabia

While sitting in the air terminal during a four-hour wait for my flight to Jiddah, I thought back on the meeting I had just attended with John Robinson, Don Aurand, Gordon Thayer, a NOAA biologist who was overseeing a sea grass study and would participate in our survey, and a number of additional scientists who would be involved with the scientific investigations during the *Mt. Mitchell* cruise. Don, Gordon, and I were still planning the details of the Leg II activities. At the meeting, I met

Dr. Ahmed al Mansi, of the National Commission for Wildlife Conservation and Development (NCWCD), who was a geologist with a degree from an institution in Great Britain. He was very interested in working with our team along the shoreline. He became an invaluable member of the team. I also met Abdul Halim for the first time. He had worked on the spill almost continuously since the oil came ashore and knew the spill area in the finest of detail. In effect, he would become our guide for the entire operation, and without his help, we could never have completed the work.

I thought further: "Despite the hectic schedule, I weathered it pretty well. But the chain smoker from Qatar sitting in the meeting next to me yesterday, the driver on the way to the helicopter this morning, and the host of smokers in this waiting area have gotten to me. I am smoked out! My sinuses burn. My esophagus burns. I'm sneezing. And I have a four-hour wait in the departure hall in Dhahran before going to Jiddah where I have a four-hour wait to go to Frankfurt where I have a four-hour wait to go to Atlanta where I have a one and a half hour wait to go to Washington, D.C. And then I die, I guess, if I haven't already before I get there."

We were flying in a helicopter past Ras Tanura at about 11:30 A.M. when I scribbled: "Turned away from the splattered yellow vomit on the floor to my right for a long view down the spit at Ras Tanura. I had been musing over whether the flock of birds on the exposed sandbar were gulls or shorebirds before being interrupted by Gordon as he was scrambling to get a plastic bag to the stricken Saudi from King Abdul Aziz University, whose white robe was white no longer. Then we headed across Tarut Bay to the military field in Dammam."

We were returning from a three-hour overflight of the spill site. One year, more or less, after the spill, there were still major sheens in the water in many places, as well as some of the most heavily oiled beaches I had ever seen.

I was grumpy about taking the flight because there were eleven passengers and only a couple of windows, so I thought we would spend most of the time looking at each other instead of the oiled shoreline. Fortunately, Dr. Kassad, a Saudi algae expert, and I were the bosses of the flight, so I got to go wherever I wanted to and saw everything that interested me, despite Kassad's passion for unoiled tidal flats. Plenty of oil was still along the tidal flats back in the bays, and a couple of the oiled tidal channels reminded me of the oiled marshes at Punta Espora, Chile, after the *Metula* spill. The overflight provided good background for preparing a work study plan for Leg II, which we worked on feverishly for the next two months.

On Thursday evening, 27 February 1992, JM and I left the office early and had our "last supper" at El Chico's Mexican restaurant before I headed for Saudi Arabia again. The actual time the *Mt. Mitchell* was going to be in the Jubail area was twenty-one days, but we had planned thirty-six days of field work for the onshore crew. Therefore, Todd Montello, a grandstudent of mine from Boston University we had just hired, and I would be heading off for Saudi the next day, where we would hook up with Ahmed and Halim. Our plan was to organize the field logistics before JM, Don Aurand, and Gordon Thayer joined us for part of the shoreline survey.

Working with John Jensen of the geography department of USC-East, we used satellite imagery to locate our ground stations and to interpret the bathymetry of the offshore area, which was labeled "shallow water" on the British Admiralty charts of the area. One of our first tasks once we got in the field would be to locate the transects we had chosen and flag them so the people on the inflatables sent out from the *Mt. Mitchell* would be able to find and continue our intertidal transects into the offshore area. We were to locate the stations with a Global Positioning System (GPS), which relied on satellite vectoring.

I wrote in my diary, "When I went to the meeting in Dhahran in December, I didn't even know what a GPS was; now we own one." That, plus a whole lot of other new equipment. I bought myself a new Nikon 6006 camera, and I had packed fifty rolls of film.

When we got to the Bluff that Thursday night, JM and I watched a short video by *National Geographic* on the Gulf War spill. We both watched in silence as the massive front of oil advanced along the shore. "What a lens!" was about all I had to say.

They showed footage of sunken oil in bottom sediments in shallow water, which was breaking up into small globules and rising like hot air balloons back to the surface. After watching the film, JM pulled out a special issue by *National Geographic* on the spill that included an artist's rendition of the same process we had just watched (that is, the rising "hot air balloons"). "That's my diagram," she said, as she pointed to it.

"You mean you did that diagram before you saw this film?"

"Yep."

"Really. How did you know it would do that?"

"It just seemed reasonable," was her quick reply.

I slept fitfully that night, the long-awaited day had finally come. I couldn't remember ever putting in that much time preparing for any project. We had been responding to oil spills for seventeen years, but I was still nervous about this trip. We had an inexperienced crew, I was still in pretty bad physical shape, and there were other problems with the equipment. Everybody at RPI worked on the packing and last-minute touches the next morning before we finally left for the airport.

I called JM from the Chicago Airport at 4:50 P.M., Columbia time. She said, "Everybody's gone home. You wore them out!"

By 4 March, we were into the routine of conducting the field surveys. We slept at the Holiday Inn in Jubail and took our meals at the NCWCD facility located nearby. Todd Baxter, a NOAA corps officer, was in charge of the logistics for Leg II, and we kept him jumping for the first few days. We had two four-wheel drive vehicles, a small boat, and other gear, and we were only running a day or two behind schedule. Ahmed and Halim were turning out to be tremendous field associates, some of the best I had ever worked with.

As fate would have it, our field survey was timed to overlap Ramadan, the fasting month for Muslims. This meant that Halim and Ahmed could not eat or drink between sunrise and sunset, which made for some interesting field time. Todd and I, and the other infidels that went out with my field team, refrained from eating any food during the day also, but I must admit, I did sneak behind the vehicle and drink some water now and then. Fortunately, we were working in the winter, and the weather was quite cool. In fact, it rained on us several days.

The first day of Ramadan was 4 March, and we surveyed three beach stations on Abu Ali, a hook-shaped headland that shelters the Jubail harbor area from the northwest winds. According to our interpretation of the satellite imagery, little of the Gulf War oil spill had impacted the Abu Ali beaches. Yet, when we got there, the entire fifteen miles of the outer beach was coated with oil from the high-tide line to the lowest of low water. The oiled surface sediment layer was about five or six inches thick, and it was a solid, hard asphalt pavement.

We were surprised to find oil there, to say the least, but we later learned that particular oil was derived from the *Nowruz* oil spill of 1983, the third largest oil spill in history. That spill was the result of the Iraq-Iran conflict, not the Iraq-Kuwait conflict. So nine years after the *Nowruz* spill, the intertidal zone of the outer beach was still solid asphalt pavement. I guess you could say that this was the first major discovery of our project.

As we were driving between two of the stations during that survey of Abu Ali, I thought, "Well, I guess the scare about mines from the war still being on the beaches was ill founded. We probably will forget all about it."

A little later at the next station, as we were aligning the stakes, a

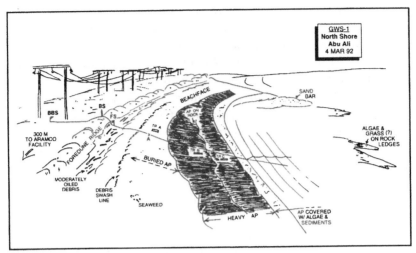

FIGURE 34: Nine-year-old asphalt pavement on beach at Abu Ali resulting from the Nowruz spill. Observation made during study of Gulf War spill in 1992.

couple of Saudi coast guard personnel drove up and talked with Halim and Ahmed. Afterward, Ahmed told us that earlier a Philippino was out on the beach where we were, beachcombing I presume, and found something interesting that he was fooling with. Then there was a big explosion, and the Philippino had not been found since. That night, Baxter gave Todd and me a little lecture on just exactly what mines look like. We did not want to become a part of the "mysterious case of the disappearing Philippino."

My biggest hassle of the whole deal was all the visitors that went out with us almost every day, fouling up the logistics I had planned the night before. When things got too busy or crowded on the ship, Baxter could always say, "Why don't we send them out with Miles?" And I did my best to give them something meaningful to do and otherwise entertain them, but it wasn't easy. Worst of all was the "film crew from hell" that about drove me crazy until I took them out on one of the big mudflats where they got bogged down in the soft mud and finally concluded they had had about enough. After that, they stayed on the ship. We still show up in quite a few film clips on *National Geographic* programs and the Discovery channel, digging holes in the mud and watching oil pour out of the crab burrows.

About a week or so into Ramadan, JM, Don, and Gordon arrived, and we worked out the details of how to describe the oiling and how to take the chemical samples. On one of the trips, we had three vehicles full of people and I was worried as to how having all those folks out in the field would work out, especially since we were still experimenting with our sampling technique. The survey went extremely well all day, but the other team finished up before us. Thus, as usual, we had too many people with us, and several of them were standing around with nothing to do as we finished describing the trenches and making the final field sketch at our last station. I decided to send Gordon's truck and the vehicle that Todd called the "kidney breaker" back to Jubail. This would also allow Halim, who was getting almost no sleep because his family slept during the day and celebrated Ramadan at night, to get home to Dammam a little earlier. Since it was the third time we had been to that station, I didn't think anyone would have a problem finding their way out. Ahmed, Todd, JM, Don, and I were left behind with a black Toyota Landcruiser.

As we loaded the Toyota for the trip back to Jubail, I asked Ahmed to drive for some strange reason. By the way, his name is pronounced "Achh' med," the first syllable being kind of like a gentle cough in the back of the throat.

I rode in front with Ahmed, and I thought he was acting a little strangely, not following the track we came in on, for example, but I didn't say anything. As we were driving along, he suddenly turned right across the middle of a small sabhka.

"No, no," I cautioned, afraid we might get stuck.

"Shortcut," he laughed.

I was beginning to get alarmed, but then I saw the rise of the elevated, heavily used track that we came in on straight ahead. I breathed a sigh of relief, waving with my left hand for him to go *shamal* (left) down the heavily used track.

Instead, he gunned it up and over the track and headed straight for the gray mud in the center of a low, linear sabhka that paralleled the track.

"No, Ahmed, No!" we all yelled, more or less in unison.

"Stop! Stop!" I yelled again.

But, no, even after we crossed over the semi-hard area by the track and out into the soft ooze, he gunned it even harder. The mud flew up over the top of the roof, and we continued to slow down and dig in as the wheels spun wildly, finally coming completely to a halt.

"Why in God's name did you do that?" I blurted out at him, knowing that he had just ruined what had thus far been an almost perfect field day. He laughed nervously.

Swearing profusely, I jumped out of the front seat and slammed the door, which closed on my seat belt which was still dangling from my right arm. Don, who was somewhat amused at my temper tantrum, particularly the difficulty I was having getting free of my seat belt, climbed out of the back seat and said, "Oh, we'll get out."

"You don't get out when you're stuck in this excrement," I responded hotly.

We spent about forty-five minutes digging, collecting boards, pushing, and grunting. All we succeeded in doing was getting stuck even more deeply. We changed our strategy. Someone would have to walk to the coast guard station with Ahmed. We could see the station in the distance, and it appeared to be about two or three miles away (it was actually six). I was doubtful that they would drive back across the sabhka in the dark, so JM and I were preparing ourselves mentally to spend the night in the Toyota.

Suddenly, a coast guard vehicle appeared on the small dune area on the other side of the sabhka. They backed up and drove around to us on the excellent track we should have taken. "Why in God's name did you do that?" was more or less what the two coasties asked Ahmed (in Arabic) when they pulled up.

We rocked the car and pushed again, and one of the coasties succeeded in driving it even more deeply into the mud, if that was possible. We would need a wrecker and a chain to get it out. It was 5:30 P.M. by then. They said they would take us to Jubail after they "broke fast" back at the coast guard station, which sat on a prominent hill overlooking sabhka world. The sun would set soon. Breaking fast time was 5:45 P.M., sunset.

We all climbed into their small Toyota, with the two coasties and Ahmed up front and the four of us—Don, JM, Todd, and yours truly—jammed into the small rear cab. Our limbs were all entangled, and our gear was piled all over us.

The six-mile ride to the station was a rough one because we sat directly on metal, and they drove at top speed along the side of the pipeline over the washboard-textured sand pile. My bottom was bouncing off the metal, the greatest insult it had suffered since the burn in the airplane crash.

"Holy hemorrhoids!" shouted Todd, and we all laughed.

While they broke fast, we sat in an open tent on the hillside beside the station drinking tea and watching the day end. Ahmed had refused food because he told us they couldn't afford to give it away. It was very peaceful in the tent. Later, they brought out some food, which we sampled sparingly. A breeze was blowing through the tent flap, and we were telling stories, very relieved for having been rescued. It was getting cool. We would have frozen our tails off if we had had to stay in the Landcruiser all night.

About dark, we climbed back into our positions in the small Toyota. An additional passenger rode in the back with us. He was a local camel herder, I guess, getting a free ride to Jubail. That gentleman was also apparently bath fasting for Ramadan, because when the breeze was just right, his body odor took your breath away.

It was a memorable trip back, with Ahmed sitting in the middle of the front seat, both arms around the small coast guardsmen, all of them shouting back and forth in Arabic, all seeming to talk at once, while the rest of us did the best we could to protect our olfactory and hemorrhoidal tissues.

We pulled into the NCWCD dining hall, which we called "Fred's Diner" in honor of the Philippino cook named Fred who made excellent dinners, just at 7:00 P.M. Gordon had done exactly what I had hoped he wouldn't—notified Baxter who flipped out, started throwing things around his room, and shouting about safety procedures, two vehicles, etc.

After dinner, we went over to the Holiday Inn. John Robinson, who had just arrived that day, was waiting for us. The first words out his mouth were, "You can always depend on Miles Hayes to disregard all safety procedures." I told him to kiss my sore ass, or something like that.

Actually, I was kind of glad to see him. I had been sending him progress reports on our findings almost every day, and I wanted to go into more detail on the oiling, doubting if anybody, including him, would believe it was as bad as we were saying. The most striking observation we made was of the heavily oiled areas on the mud flats at the heads of the bays. Even a year later, we could dig trenches in the mud, and oil would pour out of the crab burrows. We took several samples of this liquid oil, and the chemical analysis conducted later showed that it was hardly weathered at all in the year since it was spilled. This type of oiling was fairly common, and some of the oiled flats were hundreds of yards wide. And the oil just went on and on. Obviously, nature was not doing a very good job of cleaning up those sheltered mud flats. We predicted that they would be oiled for decades.

Quite a few heavily oiled halophyte marshes were in the area, and the ones on our transects were mostly dead. We could survey and dig trenches for hours without seeing any living creature. Once you got beyond the oil lower in the intertidal zone, however, polychaetes, snails, etc., were abundant in the unoiled sediments.

We saw several of our "rules of oil-spill behavior" broken at that spill. One of those rules was that oil does not penetrate sand beaches more that a few inches. However, we saw oiling to depths of over a foot and a half that had obviously penetrated down from the surface. Our explanation was that the oiled sand was "bubble sand," which has a very high porosity because of entrapped air in the sand. "Bubble sand" looks like a sponge, which it had certainly acted like with that heavy oiling.

Some days it was very depressing to see the oiling just seem to get worse and worse as we went farther back into the bays. At one of the muddy sites, we had been working for about three hours without seeing anything alive along the transect. Suddenly, a dragonfly landed on Halim's arm. We both smiled and pointed at it. He looked over at me and said, "That's a good sign. Maybe some day the life will come back here."

On 20 March, we were getting ready to escape the crowds at the Holiday Inn and move to a rather remote Aramco field facility at Tanaqib, where we would survey the northern portion of the study area. As we drove along on the super highway, I was writing in my diary about a typical day on the project, which started at 5:30 A.M. and ended around 11:00 P.M. The days out in the field usually consisted of long drives across the seemingly endless sabhkas looking for the transect locations, checking the GPS locations, and running the transect lines, all of which measured in hundreds of yards. In total, we dug over a thousand trenches in order to describe the subsurface oil. The NOAA survey team that was normally stationed on the *Mt. Mitchell* joined our field team part of the time and carried out important surveys for us.

At times the work was exhilarating. We made new discoveries about oil/sediment interactions every day. Also, it was a new environmental setting in my experience, which steepened the learning curve.

"This is great!" I had exclaimed to Todd near the end of the transect we had surveyed the day before. I was standing at the edge of a channel dominated by ebb-tidal currents, the current ripping by my legs. I had just looked across the bar to the flood-dominated channel on the other side. I suddenly understood how the oil/sediment transport system worked. The sun was out, and waves of sand were emerging along the side of the channel. The strong breeze felt good on my face. I was glad to be there.

One of our better days in the northern area (Tanaqib) was 23 March. The *Mt. Mitchell* had moved north, and on this day we were supposed to rendezvous with the teams in the inflatables as they surveyed their lines off the transects we had established. I also had hopes of seeing JM, whom I hadn't seen or talked with for weeks.

We had a bit of a flail early on. It was impressive to listen in on all the radio checks from the boats scattered all over the place. JM's boat missed the target because their GPS was not functioning properly, and we talked them in—sort of. She finally brought her inflatable to the shore, and I waded out to meet it. I held onto the side of the boat, and she held onto my arm while we exchanged information. My waders were leaking, but I ignored that. She said they had been finding some bottom oil in that area, which was unusual. One of the goals of her study was to determine if any of the Gulf War oil had sunk to the bottom during the spill. Very little had sunk, for the most part.

I thought she looked a little . . . well, forlorn, but she sounded pretty bossy on the radio, so I assumed that she was okay. We broke off the short meeting after about ten minutes, each going our own way to continue the surveys.

On that day we completed three transects, which finished our work in the northern area. At the science meeting that night in my room at the Aramco quarters, I went over our accomplishments to date. We were right on schedule and were feeling pretty good about ourselves. Everybody's eyes were sparkling with pride as we went over the data. I said, "I think we may be over the hump."

Next day, it was back to the circus in the parking lot at the Holiday Inn.

On April Fool's day, the *Mt. Mitchell* came into port in Jubail, and JM and I reunited at the Holiday Inn. I took her back to the ship at about 12:30 A.M. This day had been historic because the space shuttle had passed over the Gulf and the woman astronaut on board had talked by shortwave radio with the woman chief scientist on the *Mt. Mitchell*.

As we sat in the ship's galley eating some mango ice cream, I asked, "What did you say to Kathy Sullivan[38] today?"

"The only intelligent comment heard in the three conversations between the ship and the shuttle."

"What was that?"

"Can you see any oil?"

"And what was her response?"

"hakcoffhakcoddsspptruu bop kkksputttrereresputtterere blop!"

"I see, not a very satisfactory conversation, eh?"

"Nope."

Todd and I left Saudi Arabia on 5 April, and JM followed in a couple of days. At 11:30 P.M., I was sitting in the Dhahran Airport undergoing a massive post-spill grand blue funk while trying to write something in my diary. We had in our possession six large pieces of luggage sitting on two carts which contained the following, among other things:

1. Over two hundred sediment samples

2. Twenty-four jars of oiled sediments (JM would bring the rest of the more than 150 samples we collected.)

3. An "underground" aerial photo of the study area

4. My unused wet suit

As I looked at the pile, I asked myself how it had gone and then wrote the following in my diary:

> Actually, pretty good. As I said in my report, we accomplished all of our goals. We learned a lot more that I expected to. We worked thirty-four days without a break and thirty days without any lunch. Some days were pure drudgery and some were quite exhilarating. I tried to ignore the tragedy of the devastated upper and middle intertidal zone and focus on the uniqueness of what we were learning. Having JM in the area was nice and the three nights we spent together were outstanding.
>
> Yesterday we surveyed our last intertidal station at Gurmah Island. It was quite interesting with some oiled mangroves among several small tidal channels, but Todd was grumpy, my energy level was low, and we had the usual circus tagging along. We finished about 2:00 P.M. and were walking back to the point at the dredged channel where Pete was minding the "tin boat."
>
> I lagged behind a little, turned, and looked back down toward the profile into the sun. The tide was rising across the sand flat, having already reached the front stake with its frayed piece of signal cloth waving a brilliant orange color against the green background of the water farther out in the bay. An elevated sand bar crossed the flat obliquely between me and the flag. The sun reflected in a myriad of tiny sparkles off the partially wet surface of the spotty brown oiled sand layer on top of the bar. I studied the scene for a while, wondering if I would ever see the place again. The breeze felt good on my skin, and I felt the same sense of accomplishment that I always feel when we have completed one of those long transects. Then I felt a little sad to be leaving. I thought, "That's a good angle, maybe I should take one last picture."
>
> But remembering that I had already taken over two rolls of film, or about seventy-five pictures, of the place, I turned and walked slowly to where the others were waiting by the boat. It was over. Now all we had to do was pack and go home.

Back at the Holiday Inn, we told everybody good-bye. Abdul Halim said that it had been an "honor" to be a part of the field team. I thanked him for working through Ramadan and said we could never have done it without him.

JM, Don Aurand, Todd, Halim, Ahmed, and I returned to Gurmah Island a year later in March of 1993, once again under the sponsorship of MSRC. We resurveyed most of the stations in the southern area, noting little change. We still found liquid oil in the burrows, and some of the surface oiled sand layers were turning into asphalt pavement. When we surveyed station VII, an area that I had not thought to be very interesting during the first survey, we found that the entire three-hundred-yard-wide oiled zone was turning into a pavement, with no natural degradation of the oil in evidence. I took a picture of the surface of the flat with Don Aurand serving as a scale in the middle distance. Then I asked, "Well Don, what do you think?"

"It isn't as bad as it was last year—it's worse," was his terse reply.

Now, in March 1997, we were back to check out the area six years after the spill. After crashing Friday night and Saturday morning, we met the Aramco people Saturday afternoon who would be going into the field with us. We also had an invitation to join Aziz, Halim, Khalid, and other MEPA personnel for an early dinner at the MEPA facility, where Aziz had fixed up a mobile home behind his office as a sort of gentlemen's club— sans alcohol, of course.

Khalid cooked the dinner, which was excellent. Afterward we sat around telling stories. I told them about the Chinese scientists who visited us at USC-East after the cultural revolution was over. I explained how I had showed the Chinese my marvelous historical data on the South Carolina coast that dated all the way back to 1696, and then they showed me their marvelous historical data on the Chinese coast that dated all the way back to 200 BC. Boy, was I embarrassed. Then I told them about Deion threatening to commit *hara kiri* during my visit to the West Sea Oceanographic Institute in North Korea, a story to be unveiled in the next chapter.

During this 1997 visit to Saudi Arabia, we resurveyed ten of our stations, and as far as we could see, almost no change had occurred since our last survey in 1993. We even found liquid oil in some of the crab burrows. If anything, the halophyte marshes looked worse than ever.

On the last morning, we met Friedhelm Krupp for a few minutes at the NCWCD facility in Jubail and asked him where all of the recovery that his associates were talking about was taking place.

"In the more exposed habitats. Most of your stations are in the sheltered bays."

Okay, I guess I can buy that. After all, that was the fundamental principal of our ESI mapping scheme.

On Wednesday evening, Abdul Halim drove us past Khafji, site of the first battle in Desert Storm, and on to the Kuwait border, where some folks from Mohammad's EPA were supposed to meet us. We were scheduled to teach a course on oil-spill contingency planning and response in Kuwait, under the EPA's sponsorship, the following week.

We were delayed at the Saudi side of the border where the customs agent insisted that JM go to the women's side so the female customs agent could lift her veil to see if her face matched the picture on her passport. The mere fact that she wasn't wearing a veil didn't seem to enter into the equation. The customs agent made Halim mad, so he went looking for an officer to pass her through, which eventually occurred.

Teaching the course in Kuwait was tough, because although everybody spoke English, they still didn't understand the subtleties of our language. Worst of all, they didn't laugh at my jokes, which made for a really long week. The subtle differences between their culture and ours made it difficult to run the course the way we normally do, which was also frustrating.

We attended a dinner the last night of the course in the Sheraton Hotel, at which Mohammad and I gave certificates to the course attendees. On the way to the dinner, Mohammad dropped us off at one of his four offices to wait while he fetched his wife. We were cornered by a reporter from the *Arab Times* at Mohammad's office.

The reporter was as disappointed to see us as we were to see her because she had been waiting for a long time to talk with Dr. Al-Sarawi (Mohammad). She wanted to query him about the potential danger of the three sunken tankers near the Iraq border with their hundred thousand tons of submerged oil. She was concerned that these tankers, which were sunk during the Gulf War, would one day rupture as a result of aging, causing an oil spill of major proportions in the Kuwait area again. She was happier when she found that we were *bona fide* oil-spill experts. However, we didn't have all that much encouragement to offer about the sunken tankers.

I cringed when JM told her that the Kuwait Oil-Spill Contingency Plan was inadequate for such a catastrophe, imagining tomorrow's headline. Then I thought, "But we will be gone by then." And on second thought, "Will they ever invite us back?"

The banquet was great, and everybody thanked us for telling them about oil spills, how the world was made, the chemistry of oil, and so on. I felt a little better about the whole thing after that.

Next morning, we caught an Alitalia flight to Rome and then a Delta flight from Rome to New York. As we were flying over France, I looked down on the Gironde River and thought, "When the time comes that all of the rivers are nothing but running sewers. When we have attacked and attacked and beat Mother Earth into submission. Even the desert of northeast Saudi Arabia hasn't escaped the onslaught of mankind, as we bleed out the last of the ground water to grow wheat. What then?"

A little later we flew over the *Amoco Cadiz* spill site along the coast northeast of Brest, France, nineteen years almost to the day after the spill. From thirty-five thousand feet, the shoreline looked just fine.

PART 7

DREAMERS

North Korea

Sunday, 21 February 1993 —
Great Wall Sheraton Hotel, Beijing, China

JM and I had talked a long time about one of us doing this mission for the United Nations Development Program (UNDP). Finally, we decided that I should go instead of her for a number of reasons, not the least of which was that I could afford the time right then, and she was needed for the stand-by spill response program. A considerable amount of preparation was required, but finally, I left for the five-week jaunt with a huge pile of literature on Asia and the problem of pollution in the Asian seas to entertain myself with on the twenty-six-hour ride to Tokyo.

On the way from Portland to Tokyo, I spotted a tanker headed north and wondered what it was doing out in the middle of the Pacific Ocean. Later, we flew over the ice-encased Aleutian Islands, flying the Great Circle, you know. Thus, the tanker was no doubt headed for Valdez. Sent JM a fax from the hotel in Tokyo, wondering what it would be like to be separated for five weeks.

Made it to Beijing the next afternoon after a four-hour plus flight from Tokyo. It was cloudy most of the way until we came over the mainland, where I had an excellent view of the Yellow River and the surrounding farmland. Lots of dust was in the air, and you could see it piled up on the downwind side of the huts in the small villages.

Nobody met me at the airport, and it was an unsettling experience to wander around with a ton of luggage unable to communicate with anyone at all. There should have been a sign, "no English spoken here," but there wasn't, only blank stares. Also, no yellow taxicabs were waiting to whisk me to the hotel. Instead, a bunch of what appeared to be miscellaneous private cars were milling around the entrance, with a gaggle of unofficial-looking people offering rides to the city. After checking out everything else, it appeared that I had no other choice, so I packed the luggage into the trunk of one of the cars and climbed into the back seat. The unsavory-looking driver was joined in the front seat by a person who looked like his twin brother for the ride into town. It is a long way from the airport to the heart of the city, and we drove for an eternity along a narrow country road between rows of trees that had apparently been planted for windbreaks. Only a few other cars were on the road.

I looked around at the tall trees and thought, "For sure, these guys are going to take me out on one of these back roads, knock me on the head, steal my briefcase and cash, and leave me for dead."

I kept asking how long it was going to take to get there, but they just laughed, finally reaching the edge of the city and delivering me unharmed to the Great Wall Sheraton.

I was the only American on the team of seven experts who would help the UNDP decide how to allocate some of their funds to set up a program aimed at curtailing the pollution in the East Asian Sea. There was a range of expertise on the team, with me representing the oil-spill response community. The leader of the group was Dr. Chua Thia-Eng, director of a research group in Manila, who was an expert in coastal zone management. Every country that bordered on the sea was included on the tour. I would personally visit China, North Korea, Vietnam, Thailand, and the Philippines.

Wednesday, 3 March 1993 —
Hotel Pyongyang Koryo, Pyongyang, Democratic
Peoples Republic of Korea (North Korea)

We finished in China on the following Monday night, with a final banquet bash at which we consumed an obscene number of overly fattened Peking ducks. The previous weekend, we had braved the wintry blasts and walked a mile or so along the Great Wall, a memorable experience. Overall, I thought the China visit was very successful. I learned a lot. I assumed that the Chinese would undoubtedly play an important role in the project as it came to fruition. China is booming and will be an economic force to reckon with before the beginning of the twenty-first century.

After China, we split into two teams, with Chua, myself, and Philip Tortell, an environmental scientist from New Zealand, going to North Korea and the other four heading south to warmer climes.

At the airport, we boarded a somewhat outdated Russian passenger plane for the flight over to Pyongyang, the capital of the Democratic Peoples Republic of Korea. Chua, who prided himself on his skill with the Chinese language (his parents were from China), had trouble getting the North Korean stewardess to arrange for us to sit together. She finally sent for her boss, who spoke a little English.

The boss came storming back to us, looked at Chua, and said loudly with a heavy New York accent, "What's a matter for chou?" For a minute, I thought I was back in the customs line at Kennedy Airport. Philip and I laughed and laughed, and Chua's face got redder and redder.

When we landed at the airport in Pyongyang, I looked around for other planes but didn't see any that appeared to be functional, only partially disassembled ones of the same model as the one we were riding in. Philip and I speculated that they were the source for replacement parts for our plane.

North Korea was not exactly what I expected. It is quite hilly and probably beautiful in the spring. The city of Pyongyang has two million people and is full of big buildings, just like Beijing, except in Korea most everybody was walking and very few people appeared to have a bicycle. The most impressive thing about Beijing was the thousands of bicycles parked at all the railway stations.

As we drove to the hotel, I was thinking, "I guess this is the last real holdout for a centralized economy. It doesn't appear to be working too well. Also, it was strange to be driving past the monument to the heroic Chinese soldiers who died in the Korean War. It brought back memories of George Woodby and our star halfback, Bill Hughes, my high school compatriots, who lied about their age, enlisted early, and were shot full of holes in that war."

The Hotel Koryo was a twin tower affair, about fifty stories high. As far as we could tell, we were about the only guests staying there, and the staff left most of the lights off, which made it a little scary to walk down the long, dark hallways. Each room had a wall-length mirror, which Philip speculated was a two-way with spies on the other side. I have to admit that I was a little self-conscious when I checked in the mirror each day after my shower to see how many pounds I had gained.

Once we settled in, several people were obviously assigned to look after us, in addition to the ones on the other side of the mirrors. Our female interpreter, who spoke perfect English, seemed to take a special interest in me, the only American she had ever met. You could also say that about almost everybody else in the meetings we attended.

Our interpreter was a devout Communist, and she, and everybody else, wanted to assure us that the Great Leader was very concerned that we be comfortable and have a good time during our visit. I told her about JM, and she asked if we had any children. When I told her no, she said that I should bring JM to the Peoples Republic for a vacation up north where we could visit some magic springs that assured fertility to those that swam in their waters. I told her that sounded like a good idea, and in fact, I hope we can do it someday—not for the fertility swim, heaven forbid, but for the trout fishing, which I have heard is pretty good.

The leader of our hosts (keepers?) was a man about forty-five who had a name that sounded like "Deion," which is what I called him in my diary. They never gave us cards or lists of participants, so I don't know exactly what his name was.

We spent a couple of days in long "talking" sessions, and on 5 March, we headed for the coast for a visit to the West Sea Oceanographic Institute. Our hosts showed up at the hotel at 9 A.M. with a small van, and we rode for about an hour across the very bumpy, almost empty, freeway to the port city of Namp'o.

As we left the magnificent buildings of Pyongyang, a city that was leveled by "the Americans" during the Korean War, Deion pointed out to me that the Americans had dropped "one bomb per person" when the city was wiped out. "Don't blame me Deion, I was only fifteen years old at the time," I responded, but I don't believe my friendly interpreter told him that.

As we left the city farther behind, the scene became more and more bleak. We were driving through a dense fog, so I couldn't see much. But the fields were full of people bent over working in the rice patties. As we came into Namp'o, I saw a couple of large signs across the front of two buildings, and I asked our interpreter to translate them. She said they read: "Long Live The Glorious People's Party Of Korea" and "Long Live Our Glorious Great Leader Kim Il Sung!"

Next we visited the West Sea Oceanographic Institute, which looked a whole lot like one of the "separate but equal" black high school buildings in the South in the 1950s. The floors of the building had been recently scrubbed in anticipation of our visit, unlike the institutes in China, which basically looked like somebody had been having mud fights in the halls. All of the floors and walls of the West Sea Oceanographic Institute were extremely clean; however, even all of that hard work could not hide the austerity and bleakness of the place, particularly because it was about fifty degrees Fahrenheit inside the building. We saw some old equipment and people sitting by it acting as if they were working. After we had checked out about five rooms, Deion said it was too cold and recommended that we go elsewhere to meet.

As we walked down the hall toward the exit, I said to Philip, "Some of these labs look like they belonged to Madame Curie."

"She was better equipped," said he, as we walked out into the sun, which had just come out through the fog.

Once outside, I organized everybody for a group picture. It was a very distinguished-looking group.

From there, we went to a hotel down by the water, which was also about fifty degrees, except for the one meeting room they had warmed up just for us. This meeting went roughly, I thought, because they had come prepared to give us a wish list of equipment to buy for their poorly equipped institute. Chua tried valiantly to explain that we were not in that particular mode on this project, but that we would be glad to help them prepare another proposal for the UN that would provide funds to do just that. I had the distinct impression that they were not buying Chua's pitch.

At lunch, a typical Korean meal in the same hotel, things appeared to loosen up a bit. They seemed to think it was a real novelty to be meeting with an American. I thought we got along well, especially when I told them that the apple orchards and the hills I had seen on the trip down made me homesick for western North Carolina.

At one point during the lunch, our host Deion gave us a little history lesson. It seems that the Great Leader at the time the Chinese last over-ran the country was so disconsolate at losing the war that he committed *hara-kiri.* The same thing happened when the Japanese took over the country some time around 1920; that is, the Great Leader at that time committed *hara-kiri.* Deion concluded by saying, "and when I had to show you the miserable equipment and conditions at the West Sea Oceanographic Institute today, I felt like committing *hara-kiri.*"

"Don't do it!" I exclaimed at the end of his story, but my friend the interpreter refused to make the translation, perhaps thinking Deion should go ahead and do it after making such a statement.

Overall, I thought we were getting along well with the Koreans. They made several references to their trusting us; however, they did say an unkind word or two about how the United States was blocking access to funds from UNESCO to clean up their factory pollutants.

For our next act, we visited the new "Barrage," a colossal dam five miles long across the entrance to the Taedong River estuary, which they had started in 1981 and spent five years building. It is an outrageous proj-ect that makes the original estuary, which had a twenty-foot tidal range, a freshwater lake, destroying the life-giving saltwater marshes in the pro-cess. So much for pollutant flushing.

After we admired the dam for a while, they took us to a monument nearby and showed us a video about the construction of the dam, which involved thousands of military men. We sat in silence underground watching the video, which reminded me of some of the movie footage I had seen of the construction of Hoover Dam back in the 1930s. Long lines of military men were carrying single blocks of granite to heap on the piles that made up the infrastructure of the dam.

But, alas, somewhere along the line, problems developed. Progress slowed to a halt. What to do? Nothing short of calling in the Great Leader, who came down to the site, pawed over some maps, walked out on the piles of rocks, pointed here and pointed there, and *voila*, once again the work was back on track as the men stumbled over each other to carry the rocks even faster to the dam that would soon block off the worthless estuary.

Another problem arose when they were within striking distance of the other shore. It was not such a big problem this time, though; it only required the services of the Dear Leader, the Great Leader's son, who also looked at some maps, waved his arms, and worked the necessary magic to allow the dam construction to be brought to its hurried conclusion.

And for our last act, so to speak, they took us to the circus in a big entertainment complex in Pyongyang. It was great. The female acrobats were beautiful, and the jugglers were phenomenal. This was at 5 o'clock in the afternoon, the huge auditorium was packed, and a good time was had by all.

That night we went outside the hotel for a "typical Korean dinner" on our own, where I had to draw pictures of the food so that we could place our orders. Walking back to the hotel, we were wondering out loud what the hell we were going to do with that wish list Deion had presented to us. I thought, "This is not the easiest job I ever had."

We had a lot of good stories to tell about North Korea, and it kind of made us heroes back in Manila during the wrap-up session. As a matter of fact, as I was sitting in the Narita Airport in Tokyo on my way home, writing this diary entry, headlines on the weekly magazines in the newsstand pondered whether the Great Leader was about to start a war, because North Korea had backed out of the nuclear nonproliferation treaty, had been faking air raids in Pyongyang, etc. They had also shut out foreigners. We were apparently one of the last UN delegations to visit North Korea.

So what does all of this have to do with oil spills? What I learned from the industry on that trip was that the costs of the *Exxon Valdez* cleanup had scared them excrementless. And I also learned that there was essentially no real oil-spill response capability in place anywhere to deal with large spills—just a lot of bluster.

When I asked the officials in charge if there was an oil-spill response plan in effect, the answer was usually yes.

Can you clean up a spill?

Not yet!

Is the instrument buoy system working?

Not yet!

Not yet! Not yet! Not yet! Not yet!

At home, I wrote the following to my friend Bill Baird, the coastal engineer I had been working with in Oman:

Finished the five-week stint in SE Asia in Manila on 27 March, and from there I went straight to the International Oil Spill Conference in Tampa, Florida, where RPI had an exhibit booth and JM and I gave three papers. She had to leave the conference early and respond to an oil spill (a pipeline break) in Virginia. She didn't get home until late, night before last; therefore, since my trip we have had only one peaceful night at Deep River Bluff, which is beautiful this time of year. The dogwood trees are in full bloom, and the tropical birds are returning for the nesting season. I heard my first parula warbler of the season this morning as we were leaving for work, and the whippoorwills are singing all night long. The trees in the swamp are beginning to bud, and the wild crab apples are blooming, filling the woods with their unique fragrance. In other words, it's nice to be home! And yes, we are going to the mountains this weekend where I'm sure MOH's highwater killer will be the fly that accounts for the most trout!

Jidda

We have a saying in Arabic—whoever farms a land, must some day harvest what he farmed.

Abdul Halim Al Momen, friend and colleague

Wednesday, 20 April 1994—
Jidda, Saudi Arabia

After the bellman loaded the two heavy bags in the back of the airport limousine at the Intercontinental Hotel in Jidda, Saudi Arabia, the driver fought his way out into the bumper-to-bumper Wednesday night traffic, which is our Friday night equivalent. His brakes were catching when he stopped after racing up behind every car we came to, giving me a start each time. I was going to take the 1:00 A.M. KLM flight to Amsterdam, where I would hang around that airport for several hours before boarding a Delta jet to Atlanta-Columbia.

I had arrived the previous Friday night to help Jerry Galt, Bill Lehr, and Bob Pavia of NOAA, and Osama of MEPA teach a five-day course on oil-spill response. Osama had been in the United States for NOAA training a few months earlier, which included a brief visit to RPI in South Carolina.

I had more difficulty than usual with teaching. I spoke for a couple of hours on Sunday morning and laid a big egg, as they say. I learned long ago from my first teaching mentor at UMass, Tom Rice, that if you lose them on the first day, the odds are slim indeed that you will ever recover. And, in fact, I did have a difficult time recovering, but by the time I gave the beach cycle lecture around noon on Tuesday, I was back on track. We had about twenty diehards in the class, who stuck out most of the lectures and usually arrived no more than fifteen minutes late, which is quite unusual for an international course. Four or five students in the group had first-class potential, but we learned that MEPA, which supplied the most students for the class and like most government organizations, had the usual deadwood, budget problems, rivalries, etc.

As Osama drove us back to the hotel after class the third day, I told him, "It is very common for a hard-charging young man like yourself to have problems in an organization that has too many older, tenured staff members above him who don't work very hard. They tend to get jealous of the younger guy, and this usually leads to unhappiness all around. It's the nature of a bureaucracy."

Late the next afternoon, we drove north of the city to observe the shoreline. Once there, we noted that for tens of miles they had piled rubble on top of the fabulous coral reef that formerly existed along that part of the coast so they could build amusement parks, mosques, and highways on top of it. The coral reefs on the central Red Sea coast of Saudi Arabia are some of the longest and most diverse reefs in the whole world—that is, except where they have been covered up with rubble.

"Good for the people, but not so good for the environment," mused Osama, which I assume pretty much summarizes the feelings of the educated Saudis on that subject. I was fuming about the destroyed reefs when we stopped at prayer time for Osama to visit one of the mosques built on the buried reef. We stood on the riprap at the top of the reef front and watched the sun set over the Red Sea.

"I don't blame the Saudi people," I told Jerry and Bob, "I blame the dad-burned consultants who designed this atrocity." I used a different adjective, but *dad-burned* seems more appropriate in this book. I'm sure my father would have been equally annoyed. I continued, "They're concerned about nothing but money. They have no soul, no pride in their work. The Saudis would not have done this if the consultants hadn't recommended it."

"I hate those scumbags," I muttered, as I turned away from my two associates and looked back out across the water, thinking about the removal of the beachrock from the offshore rock platform in Kuwait and how the Kuwaitis had stopped doing it when I recommended against that practice eighteen years earlier.

When we went back south, Osama drove us through the old city and showed us the street where he grew up, reminiscing about how the streets used to be sand and he knew everyone in his neighborhood. Though only in his early thirties, Osama had already experienced the future shock of the black gold revolution that had occurred in his country within his lifetime. His home neighborhood had undergone the unbridled growth so common to many of the cities in the Middle East, with his home street by then being paved and crowded with shops and high-rise buildings.

The subject of rapid change like we had seen in Osama's old neighborhood came up in our dinner conversation later that night. I said I could think of only one place I had been where time had seemed to stand still—Cordova, Alaska.

Jerry said, "They brought it on themselves. They refused to have the bridge replaced when it fell down during the Good Friday earthquake (of 1964)."

"They sure did," I thought, "way to go gang!"

Speaking of Alaska, back in my room, a highly inaccurate film about the *Exxon Valdez* spill was on TV. It would take a lot of words to point out all of the errors in that film. They ran it over and over again all the time we were there.

Next afternoon, we supervised a field exercise in the port area of Jidda, with Osama running it in Arabic. I was standing off to myself, observing the relatively smooth operation. During the middle of the exercise, an old bugger about as ancient as me wearing a baseball cap, just like me also, walked up and said with a heavy British accent, "Probably wouldn't do much with the *Esso Valdez* would it?" referring to the relatively small amount of dispersants they were spraying on the water.

"What are you doing here?" I asked abruptly.

He tried but failed to engage me in conversation, soon giving up and joining the other gringos who were huddled off to the side. Let's face it, sometimes I'm just a natural born grump, and that's all there is to it.

Battle Ground

I walked alone on a narrow, paved strip of "no man's land" in the flat valley, pulling my rocking suitcase behind me, moving toward the last small customs station. The morning sun was bright, and I could see it shining on the high mountains on both sides of the Red Sea rift zone. As I approached the little booth, a smallish young woman in a uniform stepped out in front of me.

"Am I still in Israel?" I asked.

"Do I look like a Jordanian?" she laughed and let me pass after a quick look at my passport.

After a short flail with my visa on the Jordanian side, I was soon riding in a cab toward the town of Aqaba, which sits on the Jordanian coast at the head of the eastern arm of the northern end of the Red Sea, which is known as the Gulf of Aqaba. Across the valley the huge hotels on the shore at Eilat, Israel, stood out in sharp contrast against the mixed red and tan of the volcanic rocks and limestones of the mountains in the background.

I had just spent a week in Israel, working with Bill Baird on beach erosion issues. The previous Saturday, we took a day off to visit the old city in Jerusalem. I thought about bombs as we were rafted along the narrow streets in the teeming weekend crowds, and, sure enough, the next morning two bombs killed twenty-five people in that same area, making the peace process still a far away dream.

I was in Aqaba for a quick bit of marketing, heading for a meeting with the director of the port authority, Dr. Duried Mahasneh, a graduate of Duke University. Duried told me that while I was in the air on the way to Eilat from Tel Aviv, they had suffered an earthquake that registered 5.0 on the Richter scale. It was felt all the way to Cairo. Guess that's what happens when you build your town on an active fault line.

In between sessions of talking into two telephones at once, Duried had a few words of praise for the ESI maps that NOAA/RPI had just recently produced for the upper Gulf of Aqaba under the sponsorship of the State

Department as part of the peace process agreement. I was glad he liked the maps, but that project did have its trying moments, especially the field work.

I did the field work for that project, which started with a meeting of all the parties concerned in Cairo on 20 September 1994, at which I made a brief presentation of the ESI concept. After the Cairo meeting, I went to the beautiful city of Amman, Jordan, where I hooked up with Ahmad Khattab, an engineer who had been assigned to me the whole time. Despite the fact that I was suffering the worst attack of the "touristas" I had ever endured in my entire life, which lasted for five full days, our meetings in Amman went well, and I found the sources for most of the basic biological data that we needed for the maps. The intestinal disorder I was undergoing was the result of a macho lunch with the organizers of the project in a "authentic Egyptian" restaurant in Cairo. I was so sick that I came close to calling the whole thing off and going home a couple of times. Ahmad finally took me to a pharmacy for medicine that eased the intense stomach pain but did little else.

On the morning of 25 September, Ahmad and I flew in a commercial jet to Aqaba where we started a series of meetings, beginning at the port authority. We were sitting around a table in one of the port offices, talking and meeting instead of doing the field work that I so anxiously wanted to get started, when one of the Jordanian engineers began to explain how oil exports from the port had flourished for a time, beginning during the Iraq/Iran conflict as the Iraqis trucked the oil across the hills in Jordan to Aqaba. And then, of course, all of that stopped when the Americans bombed the Iraqi trucks into oblivion during the Gulf War. "Hey, I didn't do that," I shot back. Then I mimicked blasting a truck from a plane. Everybody laughed.

"He's a funny man," Ahmad explained.

Why not? Sometimes I wanted to add a little spice to those often totally useless meetings—useless not only to me, but to everyone else involved.

I said, "Hey, let me tell you a story." I told them about the time in North Korea when Deion laid the rap on me about the American GI's leveling Pyongyang during the Korean War. "I told him I was only fifteen years old at the time," I grinned. They all laughed again.

"But you are the American," Saleh the quiet one, who was sitting close by on my left, said softly.

Next morning, we were scheduled to fly the coast in a Jordanian army helicopter that I had been promised in one of the meetings in Amman. I didn't sleep well that night, considerably apprehensive that somehow I would not be able to complete this overflight in between runs to the toilet. And, boy, was I deflated when we arrived at the airport at 7:45 A.M. to catch our 8:00 A.M. helicopter flight, and the airport was closed.

The director of the airport finally showed up, and he had plenty of time to talk with us because, what the heck, no airplanes were coming into Aqaba that day. He located the helicopter. It was still in Amman.

The helicopter finally came, two and a half hours late. It was a new French helicopter with an Alabama-trained Jordanian pilot who walked with a swagger and didn't have any trouble understanding my southern accent over the intercom. Therefore, I knew we were in good hands.

We started at the Saudi border, and I mapped the coast in detail. Then we went back to the border, opened up the door, and I took almost complete photo coverage of the coast. This all took one hour and ten minutes.

Boy, that helicopter ride was slick. The pilot could stop the helicopter on a dime, which he did each time we came close to the Israeli border. It was a perfect platform to work from, and I took some fine pictures of the exceptional coral reefs that border the coastline. It was one hour and ten minutes of pure concentration.

Back at the airport, Hamed the driver was off running errands because we originally thought the flight would take at least two or more hours. As we waited for him to return, Ahmad and I talked with the airport director, who took time out now and then to chew out one of the many people that continually came in and out of his office. He spoke a peculiar mix of English and Arabic.

Of course, the conversation eventually got around to the Israelis. The director touched on several issues, concluding that justice and right would triumph in the end, "maybe fifteen or twenty years from now." I thought it strange to see a man with hair grayer than mine there in that tormented place still believing in justice. "But what do I know, I'm just the director of this one-camel airport (or something like that in Arabic) here in the desert," he concluded.

After that, we were silent for a while. Through the heavy pall of cigarette smoke, I studied a picture on the wall behind the director of the handsome king. And then I looked at another picture of a younger Hussain and his beautiful blonde wife, H.M. Queen Noor.

As soon as Hamed returned with the car, we were off to the marine institute to talk with a young Jordanian biologist there about providing biological data for our sensitivity maps. After he agreed to do so, he showed us their saltwater aquarium, which contained some pretty amazing corals and fish. "Hardly anyone appreciates how important the corals are," he pleaded, after a brief discourse on the unique and extreme biodiversity of the Gulf of Aqaba.

"How appropriate," I thought, "A lone voice crying in the wilderness."

"Very nice institute," I concluded, as we were preparing to leave. He apologized for the meagerness of his facility, although he did not threaten to commit *hara-kiri*. "You will see significantly better aquaria over in Eilat, because they have a much larger budget than we do."

I could not properly classify a couple of small segments of the coast from the air during the overflight; therefore, after the meeting at the institute, we went to examine them from the ground. The first place was a fancy dive club where Hamed drove right down to the beach. I checked that area out from the car and said, "Now I have to go over there," pointing across the dirt "lawn" to the fence at the far border of the property. I intended to get out of the car and walk to the fence, but without a moment's hesitation, Hamed gunned the vehicle onto the lawn, missing a bikini-clad lady lying under a beach umbrella by only inches, just like he had been doing with the semis on the main road all day. Hamed pulled right up to the fence. I grabbed my map, and Ahmad and I walked down to the water, where I made a few minor changes on the map.

A loud voice called from behind me, "You cahn't park there! Get that car out of there!" An old Brit about seventy, long and scraggly with his hair waving wildly, came toward Hamed and the engineer. He continued to yell. Our two associates walked toward him, explaining "Hey, this is the great Dr. Miles, and he is doing our country a fine service by making his map," or something like that. I couldn't really hear what they said from that distance. "I don't care who he is, you cahn't park there!" I heard that okay.

I continued putting the finishing touches on my map as the old man started racing toward us, gesticulating profusely, and chanting, "I don't care who he is!" several times. As he walked up to me, I said *"No hablo inglés!"* He shook my hand and walked away muttering to himself, completely defeated. Then I pointed across the fence and said to Hamed, "Now we have to go over there." We all climbed back into the vehicle, retracing our tracks, with Hamed once again narrowly missing the toes of the young lady under the umbrella. And at the next stop, I finished my mapping, and the field study was over.

Then I had to run to the toilet for about the fiftieth time on the trip.

That night, 26 September, I was back in the Coral Beach Hotel in Aqaba writing a fax to the NOAA contract monitor, Bob Pavia. I started the fax with the words, "Mission accomplished, but what price glory now, Captain Pavia?" And concluded with, "If NOAA gives out purple hearts, I deserve one for this mission."

Fool on a Holiday

Monday, 31 October 1994, Halloween—
Between Charlotte, North Carolina, and Roanoke, Virginia

This was the beginning leg of a week-long tour of the Southeast and Midwest as a "distinguished lecturer" for the American Association of Petroleum Geologists (AAPG). Despite the objections of the old grump who is the president of RPI, I was donating four weeks of my time to the worthy cause of educating folks on the "Exceptions to the Rules of Oil-Spill Behavior: Case Studies of Major Oil Spills of the Past Twenty Years." I also had a backup lecture on the shoreline of the Georgia Bight, the large V-shaped bend in the shoreline of the southeastern United States. I was going to be on the road for two weeks in the fall semester and two weeks in the spring semester. The AAPG was making inroads into the arena of environmental geology, and somebody thought my talk might generate some interest on their behalf. The lecture series had been in existence for a long time, but this year was the first anyone had given a lecture on an environmental topic. I was going to start on Halloween Monday at Washington and Lee University and end up at Western Michigan University on Friday, with intermediate stops at Purdue University and a couple of other places during the week.

This trip started great, with an aborted first landing at the Charlotte Airport because "we were too close to traffic" and another aborted landing in Roanoke because "a small aircraft was in the area." The ceiling was very low in both places, and the last-minute adjustment of the pilot to get up and over the elevated runway on a mountaintop in Roanoke was quite scary.

After driving in the rain and fog for over an hour, I arrived at the beautiful Washington and Lee campus where I gave two lectures in rapid succession, neither of which went particularly well. I was invited for the evening to a faculty member's house where I spent three hours of pure torture listening to three faculty members and two wives talk about life and politics on campus. Their excuse for showing no particular interest in me, or the two topics I had so slavishly prepared to excite people with on the tour, was "we don't get together very often." I wouldn't have minded being ignored by my hosts if their damn dog hadn't kept jumping on me, the meal hadn't been so mediocre, and the wine hadn't been so flat. I should have recommended that they get together more often, but I was so glad to escape that I forgot to.

They put me up in an old guest house, Morris House, right in the middle of the campus, all by myself, being the only guest in the three-story building. The architecture of the building was true to its time, I guess a couple of hundred years ago, but the interior design left a little to be desired, in my expert opinion. There was no TV and no books to read, and the Braves weren't playing that time of year anyway, so I amused myself by admiring the paintings on the walls. I was particularly attracted to a painting of "Ole Jack" (Stonewall Jackson) and "Marse Robert" (Robert E. Lee) conferring in the field before one of the big battles of the Second Civil War. That painting hung right beside my bed.

Don't forget, it was Halloween night, and I thought the building I was supposed to sleep in was more than a little spooky. I woke up at 2:00 A.M. and again several times after that, because I kept hearing the ghosts of "Ole Jack" and "Marse Robert" rattling around on the two floors above me. When they came down to the room next door, I got out of bed and went looking for them, hoping to meet the famous pair I had read so much about. I sleep in the nude, so maybe my lack of attire scared them away. Soon I heard them again upstairs and checked out those two floors as well, but I never did see them.

Next morning, I left a note on the desk of Ms. Gordon, the caretaker of the old guest house, "The room was very nice, and I found it to be quite comfortable. Thank you very much. And how thoughtful and appropriate for you to have the ghosts of "Ole Jack" and "Marse Robert" pay me a visit on Halloween night!"

That Friday evening, I was sitting in a jet plane at the airport in bad weather (again) at Kalamazoo, Michigan, waiting for clearance for take-off to Chicago. Earlier, in the afternoon, I had talked on oil spills before two hundred people at Western Michigan University, but it was a mixed audience with regard to their scientific background, so I thought that maybe the talk was a little too technical. Judging your audience is the toughest part of those tours.

After the lecture, a reporter from the local press asked me, "Would you please explain why the music died at the *Exxon Valdez* oil spill?"

"In a word, no!" said I.

A little later in the afternoon, I gave the Georgia Bight talk to another sizable audience. Dr. Marion Smith, a graduate of the geology department of USC-East, gave me a rather long introduction. She had taken some of my classes when she was a graduate student. She mentioned hearing me on NPR after the airplane crash in Alaska, but she didn't say anything about how articulate I was, thank goodness. "I was very glad to learn that he was still alive," she told the seventy-five assembled faithful.[39] "So was I," I chipped in. Everybody laughed.

"He has become a living legend," so said Marion.

"Better than being a dead one," I thought.

Toward the end of November, I went out for another week, starting at the University of Arkansas and concluding at the Southern Mississippi University. Nothing very noteworthy happened on that leg. I listened to a lot of country music during the long drives between towns, and since almost everybody was interested in hearing about the airplane crash, I pondered on occasion about what the Spirit Father might have had in mind when he saved my ass. "Hopefully, it will come to me soon," I thought, as I drove through the pretty swampland between New Orleans and Hattiesburg, Mississippi.

The spring semester jaunt was a two-week tour of the West that started in mid-February. In January, I did the ESI mapping of the Israel side of the Gulf of Aqaba, which is described in the next chapter "Middle East Dreamers."

The Saturday night before departing for the spring tour, my brother Kenneth called to tell me that Uncle Luster had died. I called my cousin Glen, and we talked. I told him I understood how he felt, while thinking "how can I help you to say good-bye?" I had to juggle my schedule, but I did make Uncle Luster's wake on Monday night.

JM was at an oil spill in Russia, so she missed the wake. I was surely glad that she and I had visited Uncle Luster and Glen when we went to see my mother over Christmas. While we were there, I told Uncle Luster, "Every young boy should have an uncle like you to take him fishing." "Why, thank you!" he replied. Then we looked over some of his old fishing equipment down in his basement and said good-bye for the last time.

The morning after the wake, I caught a Delta jet from Asheville to Atlanta on my way to San Antonio, where I gave a luncheon talk on the Georgia Bight to about eighty-five old boys, including a few I went to school with at UT. My talk went well. The old boys still dug geology.

Next morning I was fog bound, again, in San Antonio, so I was a little late getting into Lawrence, Kansas. My talk at the University of Kansas was too long, I thought.

That night, I had dinner at a restaurant with the chair of the geology department, his wife, a fellow semi-famous sedimentologist, just like me!, named Paul Enos, and his wife. Paul missed my lecture, but his wife, an architect who no longer practiced, had attended a "Suzanne's Lament" show when she was an undergraduate student at a women's college in New York state. It never ceases to amaze me how many people still remember "Suzanne's Lament."

"I wish I had been at your lecture," she whispered to me as they left, so I guess I must have been rather charming at dinner, for a change. At least we didn't talk about campus politics.

Next day, I was in Denver to lecture at the Colorado School of Mines, where I had the opportunity to see Dan Domeracki, John Horne, Steve Sturm, and Patti Tate of the old RPI Colorado office. The lecture on oil spills went pretty well, and we had an interesting discussion afterwards on "science, paid liars, and videotapes." It was the first and only such discussion we had on the entire tour. Conclusion? Science and lies don't mix.

JM met me in Salt Lake City, and we went snow-shoeing in the mountains northeast of the city over the weekend. Too soon, I had to continue on the grand western tour, flying into Burbank, California, early on Monday morning.

Delos Tucker, a fellow graduate student in the geology department at UT, with whom I had communicated off and on over the years, was my host for a luncheon lecture for the local geological society. Del had taught at a small college there for years. I talked on oil spills before thirty-five to forty assembled "friends of Delos." It was a good audience; they laughed at my jokes. In the afternoon, Del took me to his club, the YMCA, where I rode the exercise bike. I was appalled at all the baggy and flabby old men who were running around in the nude.

As I walked in the nude myself toward the shower, I noted a particularly run-down old guy. What a shock it was to realize that I was looking at myself in a mirror! Right then and there, I resolved to get into better shape, and in fact, I have worked pretty hard at it ever since then.

I had dinner with Del and his self-made wealthy friends. Later, his wife, Fran, and he showed me their family photographs and her paintings with which they had decorated the walls of the house that had been their comfortable home for more than thirty years. I spent the night there, and by the way, I did approve of the interior design of that place.

While looking at the pictures, I told Fran that she "looked like a movie star" when she was in her twenties, and that certainly was true— she was a real beauty.

Del made some money in the real estate business, but he never quit practicing geology, which he was still "having fun" with in the oil-and-gas business. He and Fran travel a good deal, going on biking trips to Ireland, Europe, etc. Del and I started in the same place, two poor boys from the South working our way through our doctorates at the University of Texas, but our lives have been on anything but a similar track since then. Nonetheless, we seemed to have a lot in common, and I thoroughly enjoyed my visit with him and Fran.

The next night, I lectured on oil spills to about eighty geologists of the Ventura Basin Geological Society in the American Legion Hall in Ventura, California. That was probably my best performance to date. Doug Imperato and Vince Ramirez from the old RPI Pacific office were there, and they seemed to enjoy it.

The guy who introduced me had taken one of my AAPG field seminars. He suggested in a brief conversation before the talk that the next civil war, noting my interest in the civil wars of the South, would be over "the present onslaught on and attempt to destroy the Constitution of the United States." I wasn't sure what he was talking about until he opened his introduction of my presentation with an anecdote about his kid's first grade teacher asking her students to paint an oil spill over the cute drawings they had made of the "mountains and deer" on the North Slope of Alaska.[40] "That's what our kids are being taught about the oil industry," he concluded.

I noted a strong undercurrent of resentment among the oil industry people in that area toward environmentalists. Most of the folks in Santa Barbara prefer not to see any more offshore rigs marring their views of the fabulous sunsets and are the most vociferous in their resistance to more offshore exploration in that area.

At the end of my talk, one of the old boys in the audience asked if I was going to give that same talk over in Santa Barbara, implying that it would help to better educate people in that area on the harmlessness of the oil industry, I presume. What? After I had just told them that the second largest oil spill in history was the blowout of the Ixtoc I offshore platform in the Gulf of Mexico. If you remember, it was thirteen times as large as the *Exxon Valdez* spill. Most of the oil wells in the Santa Barbara area are offshore platforms. After fielding that question, I thought "Sometimes I wonder if anybody ever listens to anything I say."

Next day it was on to Cal State Haywood, a school that sits on top of the notorious Haywood Fault a few miles south of Oakland, California. My talk and visit were uneventful, but I was impressed by the tour they gave me of the bending sidewalks and warping houses built across the fault. That long day ended about 1:00 A.M. in the Sheraton Hotel in downtown Anchorage, Alaska.

Between eighty and ninety people turned out for my luncheon talk on oil spills at the monthly meeting of the Alaska Geological Society at the Sheraton the next day. The talk went well, they laughed at my jokes, and several good questions were raised at the end. At last! The eighteenth stop on the tour, and things finally seemed to go right.

The next morning, a Friday, I took one of the most scenic airplane rides I could remember between Anchorage and Fairbanks, flying right by Mt. McKinley and studying the spectacular stream patterns in the snow and ice the whole way.

It was bitter cold in Fairbanks, where I gave a so-so lecture to a good-sized audience at the University of Alaska, which was highlighted by my stepping too close to the edge of the stage and falling off flat on my face. And what did I say when I got up off the floor? "Next slide please."

When we took off from Fairbanks the next morning, the pilot told us that the ground temperature was -39 degrees F. Once more, I was treated to the spectacular Alaska winter scenery between Fairbanks and Anchorage. On the way south to the lower forty-eight from Anchorage, I had a very good view of Egg Island on the Copper River Delta, which was all covered with snow, and once more, I looked straight down on the site where Jamie found me lying on that rock.

This time I was headed for Los Angeles and the 1995 International Oil Spill Conference, where I was scheduled to give two talks: one on the ESI for river systems, and the other on the Gulf War spill. My talks went okay, but JM was the star of the show. She gave three talks on Wednesday morning, 1 March 1995, that had enormous, standing room–only crowds. She was breaking new ground on the topic of heavy oils that do not float, a new product that could dramatically change the way we respond to oil spills.

That night we went out with the gang, had a nice tuna dinner, and told some stories.

Now, was that a fool on a holiday or what?

Middle East Dreamers

There is a logic to peace now that cannot be denied.

Secretary of State Warren Christopher, on the
evening news on 14 January 1997

Wednesday, 22 November 1995—
Coastal North Carolina

With nothing better to do while driving from Myrtle Beach, South
Carolina, to Wilmington, North Carolina, on a marketing trip to sell our
"new initiative" on resource mapping to the U.S. Army Corps of Engineers, I listened to the Howard Stern talk show on the car radio. My
brother Edwin had just recently referred to Howard as "the lowest of the
scum," but I preferred him to the myriad of other talk show hosts I had
at my fingertips because he was the only one, as far as I could tell, who
wasn't actively promoting the overthrow of the government. As his associate, Robin, read the news, she noted that a tourist had recently died
in Eilat, Israel, during an earthquake. "Why in God's name would anyone
go to Israel for a vacation?" queried Howard. I laughed at Howard's in-
bred New York ignorance because I knew that Eilat was in fact a plush
tourist resort crowded with European tourists in the summertime.

Then, as I sped along the highways across the exquisite blackwater
rivers and swamps of coastal Carolina, I thought back to my last trip to
Israel earlier in the year.

I arrived in Tel Aviv on the afternoon of 23 January 1995 and eventually made my way to the Wind Mill Hotel in Jerusalem. I planned to be
in Jerusalem for only one day, during which I would begin gathering data
for the mapping project I was supposed to carry out in the northeastern

arm of the Red Sea (the Gulf of Aqaba) for NOAA under the sponsorship of the State Department as part of the peace accord. My job was to complete the NOAA/RPI part of the sensitivity mapping of the Gulf that I had begun on the Jordanian coast a few months earlier (discussed in a previous chapter "Battle Ground"). After Jerusalem, I planned to go to Eilat on the Gulf to begin the mapping job, and after that was finished, I would hang around there for three more days in order to attend a major regional scientific meeting on the ecology of the Gulf.

Next day, I had an appointment with Dan Perry, director-general of the Nature Reserves Authority in Israel, who, I assumed, would supply some critical data on the biology of the Gulf that we could encode on the maps. I arrived on time for the noon meeting, which wasn't too difficult because my body thought it was dinnertime, but old Dan was half an hour late. During the wait, I sat in the hall and amused myself by watching the beautiful Israeli secretaries parade in and out of the door at one end of the hall and rough-looking, sun-tanned men with forty-fives strapped to their hips parade in and out of the door on the other end of the hall. Needless to say, I was wondering what was going on behind those closed doors, and I never found out.

When Dan Perry showed up, we dashed into his office, and he started signing checks, basically ignoring the latest examples of our ESI maps that I had laid before him. It took Dan, who I thought bore a striking resemblance to Abraham Lincoln, about twenty minutes to sign all of the checks and rummage through the rest of the stuff on his desk, after which he said, "Sorry, I have another appointment at the university. I have to go now."

When we walked out of his office, I asked one of the beautiful Israeli secretaries to call me a cab. "Oh no," said Dan, "My driver can take you back to your hotel as soon as he drops me off at the university." Great! Finally, I didn't have to pay for something! Thus far, nobody had met me at the airport, I had to find my own way to the hotel, and I was paying for everything myself, unlike in Jordan, where they laid out the red carpet.

On the way to the university, Dan apologized for being so rushed, right after he had thoroughly reamed out the driver for some offense that I was unaware of. I almost recommended to Dan a good correspondence course in time management that I knew about but thought better of it. As you can probably guess by now, we had not yet managed to establish the kind of rapport that would allow suggestions for self-improvement. "We will be able to spend more time together at the meeting in Eilat," Dan said as he got out of the car at the university. I couldn't wait.

The next day, 25 January, I got up at the ungodly hour of 4:30 A.M. to take a cab to the Ben-Gurion airport in Tel Aviv to catch a plane to Eilat. Then there was the hair-raising ride in the dark down the mountain at breakneck speed. This was surprising because, in general, the driving in Israel is the tamest I've ever seen in a foreign country. I think the driver was mad at me because I was trying to bargain with him on the price of cab fare at 5 : 00 A.M. . The speedometer broke when we hit 160 km/hour on one of the steeper slopes, and from then on we were flying blind, so to speak. Trust me, we were going very fast. I was so glad to be alive when we got to the airport that I gave the driver a two dollar tip. So there!

At the airport I met Ellik Adler, my primary host and head of the Marine and Coastal Division of the Ministry of the Environment. In the mad scramble for seats on the ARKIA jet, Ellik managed to procure me a window seat. The landing strip is adjacent to the beach in Eilat. There-fore, I had an excellent view of most of the Israeli coastline of the Gulf of Aqaba as the plane was landing.

We rented a car and toured the entire Red Sea shoreline of Israel, mak-ing several stops for photographs and closer inspection. Except for the two miles of the Coral Beach Nature Reserve, which starts at the Egyp-tian border, most of the coastline is highly developed. Many large tourist hotels line the beaches, especially along the north end of the Gulf.

After the initial tour, we visited the Marine Pollution Prevention Sta-tion, which reports to Ellik in the Ministry of Environment. That unit, which has boats, boom, and other equipment adequate to deal with small oil spills, is located adjacent to the oil port. The tanker was in port when we arrived; thus, they had strung out a line of boom in a configu-ration designed to deflect any spilled oil to the shore by the port rather than let it go into the Coral Beach Nature Reserve to the south. All of the oil that is presently brought to the port is transported in by this one tanker, which runs back and forth to the oil field on the other side of the Sinai Peninsula in the Gulf of Suez. This field was discovered by the Israelis when they controlled the Sinai. The oil brought to Eilat by this tanker, which makes five or six calls per month, is sent north via pipeline.

Spills have occurred there in the past. Several of the classic papers on oil pollution of coral reefs have dealt with the chronic oil pollution of the reefs in Eilat. There was talk of an approximately eighteen-hundred-gallon oil spill of the tanker *Puma* in June 1992 that did have some im-pacts on the coral reefs down in the coral reserve.

At 6:00 P.M., I gave a spine-tingling lecture on the "Impacts of the Gulf War Oil Spill" to an audience of fifteen or so in a dive shop some-where in Eilat. Every corner in Eilat has a dive shop. I'm sure there are more SCUBA divers per gallon of water in Eilat than anywhere else on earth. Several people told me that the small reserve area averaged about two hundred thousand dives per year!

My audience was composed mostly of the marine pollution unit personnel and several people from the Coral Beach Nature Reserve. There was lots of interest in the talk because, though it happened just across the Arabian Peninsula from them, they had not heard a whole lot about "the mother of all oil spills."

Back at the pollution control unit, they had a birthday cake for one of the responders who was an ancient thirty-three on this day. I was impressed by the *esprit de corps* of that team. It was a pretty tough-looking group. If ever there were any trouble, I would have wanted them on my side.

Later, Ellik and I continued our discussions over a dinner of sesame shrimp at the Last Refuge restaurant, which overlooks a marina. All in all, 25 January 1995 was a pretty good day.

The next morning, we visited the Dolphin Reef, an area enclosed with nets, where five or six dolphins imported from Russia (?) were available for feeding displays and for the tourists to swim with. Oh joy! I made the following comment in my field notebook: "So, we will be able to show some mammals on our maps, after all." We weren't expecting to find any mammals that far north in the Gulf.

Paula Levin, the office manager, showed us the facilities before we were joined by the owner, Roni Zilber. This is a private facility about seventy-five meters wide squeezed in between the oil port and the cargo port. Roni reminisced about the 1992 spill, which messed up his nets, killed some corals, and infected the eyes of the dolphins for several days. He talked at length about the new corals growing on his nets, the "healthiest corals in all of Eilat," because he supervises the divers and doesn't let them touch the corals. Ellik said human divers are the biggest threat to the corals in the reserve. Roni asked me to get in the water with him so he could show me how healthy the corals were. I declined his invitation "for now."

Roni has a contingency plan for when, or if, the big spill at the oil terminal occurs: "I will get in the water, open the net, and then swim out to sea with the dolphins as far as we can go!" Needless to say, Roni was a pretty interesting fellow.

We had several other meetings with the local scientists during the day, including Dr. Reuven Yosef, director of the International Birdwatching Center, which is located in Eilat. He was very helpful and said he would give me information on birds sometime during the meeting next week, where he, and just about everybody else we talked with, would present a paper.

As it turns out, Eilat is one of the more famous bird-watching areas in the world, especially for raptors. Reuven told me that the main reason Eilat is such a good place to observe birds and study migration is its location at the junction of three continents, where twice a year millions of birds that breed in eastern Europe and western Asia squeeze through the Middle East on their way to and from their African winter quarters. Over four hundred bird species have been observed in the Eilat area. Unfortunately, I was there at the wrong time of the year for the best birding.

Our maps are designed to show the seasons when the most sensitive birds, such as white pelicans, waterfowl, and certain wading birds, would pass through. A few cormorants were present when I was there, and the only important nesting bird in the area is the rare (in Israel) Kentish plover, which lays its eggs in the sand by the cargo port. Plovers seem to be threatened the world over because of the loss of habitat—too many hotels! The bird data I was able to procure from Reuven turned out to be excellent.

We had a lunch meeting with Aaron Miroz, the marine curator of "Eilat's number one attraction," according to the many signs, the Coral World. This place is phenomenal, all right, with its underwater observatory in the reef and an accompanying yellow submarine named Jacqueline (there are only two more like it in the whole world). The tower that contains the underwater observatory has a circular restaurant on it that is perched about twenty feet above the water. To get to the restaurant, you walk on a pier that crosses the entire width of the fringing reef, a spectacular view. A phenomenal saltwater aquarium is in the facility on the beach, which has a wide array of live corals and fish living in it.

At 3:30 P.M., I dropped Ellik off at the airport for his trip back home. Being inspired about being in one of the best bird-watching spots in the whole world, I left the airport and headed for the Eilat Mountains Nature Reserve with my Middle East bird guide and binoculars in hand. I read in one of the bird books of a trail back there with breathtaking views of the Gulf of Aqaba from the crest. I climbed up the steep, rocky trail for forty-five minutes, finally reaching the crest where I had a breathtaking view of the sewage ponds up in the rift valley north of Eilat. Guess I was on the wrong trail, but I did see two "life birds"—the sand partridge and the white-crowned black wheatear, both of which sang beautiful melodies. I would have continued in my quest for breathtaking views of the Gulf of Aqaba, but I had to scramble and slide back down the trail to the car before dark. It got pitch black around 5:30 P.M.

On Friday, 27 January, I started at the Egyptian border and walked the entire length of the coastline to the Jordanian border, moving the car as I went along. I moved very quickly in the nature reserve, but had to slow down some in the more crowded tourist areas. I found stepping over the bare-breasted European women to be a bit of a nuisance, but I learned to live with it. It was almost dark when I got to within a couple of hundred meters of the Jordanian border but couldn't quite reach it. I was well beyond the end of the road, and ahead of me were several barbed wire fences and flags. Need I say more! Anyway, I had a pretty good oblique aerial photograph of the area, and I planned to examine the shoreline with a boat the next day, Saturday.

In addition to the boat survey on Saturday, I spent two hours at Coral World Eilat, studying the corals and fish in both the extensive aquarium and the amazing underwater observatory. One hundred species of coral and I don't know how many different species of fish are present in this part of the Gulf. Thousands of fish were within view of the underwater observatory. Exceptional *biodiversity* is one of the key attributes of the reefs in the Gulf of Aqaba.

On Monday, 30 January, I attended the opening session of the meeting on "The Ecosystem of the Gulf of Aqaba in Relation to the Enhanced Economical Development of the Peace Process—II." On that first day, about fifty people were in attendance, including several Egyptians, one Jordanian, one Moroccan, a couple of Europeans, and three or four Americans. Some of the highlights of the oral presentations were:

1. Dr. Israel Peleg, general director of the Ministry of Environment, said he thought "the era of peace is the time for the environment." He cited the State Department/NOAA ESI mapping exercise right up front and introduced me to the rest of the audience.

2. Professor Micha E. Spira, of the Interuniversity Institute for Marine Sciences of Eilat, organizer and host of this meeting, said the main theme of the meeting is to balance development with environmental preservation. He ended with a great quote, translated literally from the Hebrew, if we don't develop properly, "we will be cutting the branch upon which we sit." If ever a place was undergoing development pressures, Eilat is it!

3. Ron Pundik, introduced as the "architect of the Oslo agreement," gave a fascinating historical account on how that agreement came to be. He lamented that the "last two weeks has been the worst since September 13" (1993, when the agreement was signed at the White

House). It was an unsettling time because several Israeli soldiers had been killed just a few days earlier. He then told us how only two years ago a couple of frogmen from the Islamic Jihad of the PLO had swam ashore exactly where we were sitting one night and killed one of the university's night watchmen. The lab we were meeting in was right on the beach.

Later, Dan Perry and I discussed the preservation of mammals in the Israeli nature reserves, possibly the only place they are being protected in the Middle East. I had seen one of the famous ilexes, a large deer-like animal with long, saber-shaped horns, on one of my treks in the mountains. Dan seems like a pretty nice guy, after all (i.e., ignore everything I said on day one).

The subject of the morning session the next day was the physical oceanography of the Gulf. The speakers didn't appear to know much about it. I looked on the maps at the narrow entrance to the Gulf, the Strait of Tinan, and thought "that's where the big spill will occur."

But guess who was the star of the show on Tuesday? None other than Bullet Dan Perry! He was in the unenviable position of speaking last (at approximately 6:30 P.M.). However, without a note or slide one, he held the audience spellbound with a chain-of-consciousness speech that said:

1. The clarity of the water is the number one issue.

2. Preservation has to begin outside the protected area.

3. You have to consider the whole Gulf.

4. The port is still discharging phosphate, and the city is still discharging sewage.

5. "And it isn't if, only when, will the big oil spill come."

6. He is going to cut the number of divers in the reserve.

This speech so inspired the audience that several people proposed that the meeting pass a resolution to protect the environment of the Gulf, to be agreed to by all countries, I guess. I wondered if the approximately forty Israelis, five or six Egyptians, one Jordanian, and no Saudis present constituted a forum. The Jordanian biologist, the same one who had shown me around the institute over on the other side of the Gulf during the earlier trip, one young Dr. Salim M. Al-Moghrabi, finally stood up and said he loved the coral reefs too, but the man on the street in Aqaba (Jordan) wanted the peace process to "put more money in his pocket because he only makes about two hundred dollars a month." I guess that means there are going to be a lot more hotels in Aqaba.

Finally, the lone politician remaining in the audience, a Palestinian who was a deputy minister of agriculture said he always thought Dan was something of a horse's rear end, and had apparently told him so on more than one occasion, but that his speech was great. My sentiments exactly. The deputy minister was so inspired by Bullet Dan's speech that he said he would personally lobby for the proposed "Save the Gulf" resolution at the next meeting of the Knesset.

After all that excitement, the only thing I could think of that would calm me down was another sesame shrimp dinner and a bottle of Carmel Sauvignon Blanc at the Last Refuge restaurant.

The next day, Wednesday, 1 February 1995, the peace process meeting continued at the university. Overnight, two of the participants (one of whom was probably the Israeli H. Shural) had written a manifesto by "concerned scientists" on "saving the Gulf of Aqaba from environmental destruction." Chairman Spira put the manifesto to a vote, and most people raised their hands in support. Spira didn't seem too enthused about the idea. I don't know exactly why, but such covenants have never excited me very much either.

The thing I remember most about this day, of course, was the talk I gave at 5:30 P.M. on the sensitivity mapping project. Naturally, the chairman of this particular session was more conscious of the timing than any in the other sessions. Ellik Adler spoke just before me, and the chairman rushed him through his well thought-out presentation on the oil-spill contingency plan recently published for the Gulf of Aqaba as part of the peace process work. Ellik stuck with his planned speech and made all of his points anyway.

Meanwhile, I thought, "If he holds me to fifteen minutes, I'll never make it. I've got sixty slides in that tray."

About seventy people were there for my talk—the room was literally packed. I started out by saying that "the sensitivity mapping technique I am about to present is not the product of the meeting of a committee of bureaucrats in Washington, D.C., nor was it formulated in a test tube in the Exxon Research Lab in Houston, Texas; no, it is the product of twenty years of observing live oil spills in the field." Then I briefly introduced NOAA's spill response program and gave a quick review of some of the major spills studied since 1975.

I observed the audience carefully as I talked, speaking slowly, hoping they could understand me. When I showed pictures of the *Urquiola* spill in Spain (1976), where the ESI idea was first conceived, I heard gasps pass through the audience as each grim sight of the shoreline drenched in oil was flashed on the screen. This degree of attention followed right on up to the photographs of the oiled crab burrows at the Gulf War spill. The folks in the back had been rather boisterous throughout the day, and I wondered how they would react during my talk. The place was so quiet you could hear a pin drop.

When I came to the human use and biological resources aspect of the mapping, I briefly filled them in on the status of the work in Jordan and Israel, acknowledging some of the people in the audience who had contributed data. I even put in a plug for continuing on down the shoreline into Egypt. I was especially pleased to note that Dr. Badarvi, head of the Egyptian delegation who was sitting in the second row, nodding his head right with me as I explained the icons we use for biological elements.

The timing was good. I finished with an example of the application of the ESI during a spill.

When I showed the last slide and turned back to the audience, there was a moment of silence. Then from the middle of the room, one of the Israelis, I think probably the one that wrote the manifesto, said in a moderate voice "Bravo!"

Then the applause set in.

It was one of the proudest moments of my professional life. The years of observation and effort had been crystallized into fifteen minutes for an audience of seventy mostly hard-core scientists who were poised to do something, anything, to contribute to the most likely lost cause of saving the uniqueness so peculiar to the Gulf of Aqaba. I think this crossing of the old lines of division to a certain single mindedness, even if the participants from the Arab side were few in number, is what the peace process is all about.

Guess I'm as bad as the "concerned scientists" after all.

Ensoulment and Restoration

One does not sell the land upon which the people walk.

Crazy Horse

Friday, 4 April 1997—
Deep River Bluff, Calhoun County, South Carolina

As I sit at my desk on a beautiful spring morning at the Bluff, I can hear a red-shouldered hawk calling down over the swamp and a red-eyed vireo, a tufted titmouse, and a pine warbler singing in the trees limbs over the upper deck. I'm trying to think of a way to bring some optimism into this story about the black tides I have observed all over the world. Even though the black tides that happen all the time are some of the least of our worries when it comes to the destruction of the earth's natural systems, I do believe they are symbolic of the greater danger from more lethal insults, which is why I have described them in such laborious detail in this book.

In my readings about the native peoples of Alaska, I recently came upon a short article entitled "Ensoulment of Nature" by Gregory A. Cajete published in a book *Native Heritage*.[41] In that article, Cajete says that the importance the American Indians put on connecting with their place of origin (i.e., the environment) is not just a romantic notion out of step with the times, but rather "the quintessential ecological mandate of our times."

He said the Indian people experienced nature "as a part of themselves and themselves as a part of nature." In effect, they saw no separation between humans and the environment. Early Indian practices appeared to be founded on the reality that the environment must be sustained in order for the people to be sustained. Ensoulment means that the human participates with the earth as if it were a living soul. Not being an expert in this area, I hope I've stated his concept correctly. If we all adopted this frame of mind, I'm sure it would be one more step in the honorable direction toward making the earth a better place to be, which is something all of us tree-huggers and right-thinking engineers want to do, correct?

But what about that part of the natural system we have already messed up? One of the things we do in our business is work on restoration projects, primarily those dealing with oil spills. Consequently, we are familiar with the rather extensive literature that exists on the subject of restoration of natural systems (e.g., Committee on Restoration of Aquatic Ecosystems, 1992;[42] Thayer, 1992[43]). Many restoration projects have been attempted to date, and some are notable successes.

Not only are the abuses halted in restoration, but the system itself is physically restored, and possibly some missing components are replaced. The system probably never recovers to its natural state, at least not in our lifetime, but it is returned to at least a partially functioning condition. Working in our favor is the phenomenal resiliency in the mechanisms of the earth. As Rene Dubois pointed out, "a river or lake is almost never dead. If you give it the slightest chance by stopping pollutants from going into it, then nature usually comes back."[44]

One of the most striking successes with restoration in recent years has been the recovery of many of our rivers after passage of the Clean Water Act of 1970. Even some of the rivers that ran black in western North Carolina, the ones that so upset my father, are now relatively clean. Another example is that the prevention of the sale of DDT in the United States has allowed for a phenomenal recovery of the brown pelicans along the South Carolina and Louisiana coasts. I am also a nonparticipating member of Trout Unlimited, an organization that has restored many trout streams with the assistance of voluntary weekend workers.

Earlier this year, as part of one of our ESI projects, I mapped the headwaters of the St. Johns River in Florida, where an enlightened water management plan has overcome the traditional flood control projects of the 1950s and 1960s to preserve more than 125,000 acres of pristine and restored freshwater marshes. Believe me, it is a phenomenal natural area with nesting bald eagles and thousands of other birds in the resplendent cypress-tupelo swamps and marshes.

In conclusion, it would seem that the answer to the primal question is that first we must once again become as one with the earth, following the guidelines of the gatherers and leavers, such as Tecumseh, Crazy Horse, Chief Joseph, and the rest, preserving as much as possible of that which has not been irrevocably changed. Then we have to restore what we can to as close to the original form as possible. To do this, needless to say, the population explosion must be curtailed.

Yes, I do believe that restoring the earth is our one last great hope, and, obviously, so do a lot of other people, considering all the restoration projects taking place at this very moment.

The old men
say
the earth
only
endures.
You spoke
truly.
You are right.
The Earth Only Endures

Bureau of American
Ethnology Collection
(Brown, 1970)[45]

Afterword: My Oil-Spill Heroes

> Strong hearts to the front,
> Weak hearts to the rear.

_____

Crazy Horse

Sometimes responding to oil spills can be pretty tough work, requiring sacrifices of big chunks of time away from home and family, putting in very long days, and doing sometimes risky and dirty work. The business is definitely not for the faint of heart. Over the last twenty-three years that we have been in this business, I have observed many people at work at spills, and in this Afterword, I will list a few of my very own personal oil-spill heroes who have gone more than that extra mile in the discharge of their duties.

My number one heroes are:

1. JACQUELINE MICHEL (JM): I doubt if another scientist in the world has the depth of understanding of this topic that she does. Not only is she excellent in the field and understands the physical and biological processes at work during a spill, but she is a good chemist as well, which makes her the most versatile spill responder I know. She has written a number of manuals for NOAA that cover a broad range of topics from ESI mapping to _in situ_ burning to mechanical protection measures. However, she is at her best when confronted with a whole new set of conditions when the next spill occurs, coming up with answers and recommendations within minutes after the phone call. The RPI spill response program would have failed long ago without JM at the helm.

2. ERICH GUNDLACH: Erich was with me at my first oil spill in Chile and eventually took over the oil-spill response program at RPI when I became too involved with other things. He is one of the hardest workers I have ever met. His major strength when he was with us, besides being a good scientist, was the ability to confront almost any

situation in the field and still rough things through to a satisfactory conclusion. He was also excellent at seeing publications through to the end. If it hadn't been for Erich, I'm sure our program would never have gotten the start it had and may not have made it at all.

3. ED OWENS: Ed was one of the most productive graduate students I ever had. His outstanding professional career speaks for itself. Ed was perhaps the very first coastal geologist to tackle the oil-spill issue; he certainly preceded me. In fact, when he came to UMass and wanted to discuss his work on spills, I showed no interest whatsoever. He was always great at organizing complex projects and managing groups of people. His work organizing the *Exxon Valdez* scientific response, the SCAT surveys in particular, is a legend in itself, as far as I am concerned.

4. JOHN ROBINSON: John is an engineer who got his start with NASA. He was always an unusually creative and innovative manager. Not only that, he had such an engaging management style that the people who worked for him were totally devoted to his vision. *Vision*, that's the word that best fits John Robinson and his contribution to the oil-spill response business. He had the vision to start and nourish NOAA HAZMAT, the most experienced and effective scientific spill response unit in the country and, I assume, in the world. He was also responsible for the vision and management of the NOAA *Mt. Mitchell* cruise to the Arabian Gulf, which was one of a kind.

5. JERRY GALT: If an oil spill ever occurs in U.S. waters, Jerry Galt, or one of the members of his team, will be the first people called to forecast the trajectory of the oil. He has responded to a phenomenal number of spills, and nobody else comes close to having his understanding of all the vagaries of oil behavior on the water. He has created a program that sets the standard for the field. All of this is great, but what I like most about Jerry is his ability to think up new ideas. He is a very creative guy. Every time I hear him give a lecture, I am expecting to learn something new, and I usually do. Also, he comes up with great sayings like "post-spill grand blue funk."

6. DAVE KENNEDY: I first met Dave in Alaska in 1976, and we started talking about the *Metula* spill. Little did we know that we would soon be on John's team in far away places like Brest, France, at the *Amoco Cadiz* spill and be spending Christmas together in Puerto Rico fending off fuzzy-headed scientists. Like John and Jerry,

Dave has been to a lot of spills. Dave, like Erich, but with a different style, had a dogged determination to get the job done, and I'm sure John would say that the NOAA HAZMAT team would never have happened without Dave being there. He was a great team player in the early days, and now he is the boss. That's the way it should be, isn't it?

Notes

1. Comet, P. A., "Geological Reasoning: Geology as an Interpretive and Historical Science: Discussion," *Geol. Soc. Amer. Bulletin* 108, no. 11, 1996.
2. This type of high-wing Cessna aircraft has fuel tanks inside both wings, near the fuselage of the plane. I assume the tank in the left wing ruptured when the wing broke off.
3. This is the most critical time during a crisis like this one. It is called the *recognition*, or *orientation*, phase. One has to first come to grips with what has actually happened before the proper remedial action can be taken. I learned that in the Exxon-sponsored HAZWOPER training session that I took in preparation for the *Exxon Valdez* field work, but I had not yet begun to rerun that mental tape when I made these particular comments.
4. I have tried many times to remember exactly how I managed to get out of the plane and stand on the pontoon with my legs in that condition. Some people have told me that I can't remember it because it was too painful. I don't know.
5. Typical water temperatures in these waters in late summer are around fifty degrees Fahrenheit. I have no corroborating data on what the water temperature actually was in the bay where we crashed. On this particular day, the breeze produced a wind chill that no doubt contributed to the potential danger of hypothermia.
6. I had forgotten at that time that Todd had called in our position to the flight service a few minutes before the crash. Therefore, JM and her pilot would have had a location to start the search and not have to start it all the way back at Petersburg.
7. The average oil tankers in those times could transport around ten thousand tons of oil. The *Exxon Valdez* carried two hundred thousand tons. The largest modern supertanker carries over three hundred thousand tons.
8. "Impact of Oil Spillage from World War II Tanker Sinkings," report no. MITSG 77-4 (MIT, 1977).
9. Etkin, D. S., "International Oil Spill Statistics: Oil Spill Intelligence Report," Cutter Information Corporation (Arlington, Mass., 1996).
10. Etkin, D. S., "The Financial Costs of Oil Spills: Oil Spill Intelligence Report," Cutter Information Corporation (Arlington, Mass., 1994).
11. Galt, Jerry, and Gary Shigenaka, unpublished trip report for NOAA HAZMAT (the Hazardous Materials Response and Assessment Division of NOAA) (Seattle, 1997).
12. Ellis, Jack, *The Sunfishes* (New York: Lyons and Burford, 1955).
13. From "A Child of the King" by H. Buell and J. Sumner.
14. Green, C. H., *Birds of the South* (Chapel Hill: University of North Carolina Press, 1933).

15. Bergman, Ray, *Trout* (New York: A.A. Knopf, 1947).
16. Hayes, Miles O., Ed Owens, Dennis Hubbard, and Ralph Abele, "The Investigation of Form and Process in the Coastal Zone," paper presented at the annual meeting of the Geomorphology Symposium, Binghamton, N.Y., September 1972.
17. Quinn, Daniel, *The Story of B* (New York: Bantam Books, 1976) p. 278.
18. During the 1969–1971 surveys in Alaska, I saw very little of Prince William Sound, the site of the famous *Exxon Valdez* oil spill. I wrote this description of the Sound in 1975, but it was not until I worked on the spill itself that I truly experienced its uniqueness and beauty. It seemed as if the theretofore unknown blue world was reaching out to me in a special way on this day.
19. I also took this slide show on a few road trips, but it never achieved near the popularity of "Suzanne's Lament." The message was not as clear. One viewer at the University of Wisconsin said it was more "light hearted" and "fun" than "Suzanne's Lament." In other words, it didn't make anybody cry—except me.
20. Toward the end of May in Yakutat, we met George Ochenski, a mountain climber who loaned me the book *Cold Mountain*, one hundred poems by Han-Shan, a Chinese Buddhist poet-recluse who lived nine hundred years ago. Han-Shan, *Cold Mountain* (New York: Columbia University Press, 1970).
21. The "blue" or "glacier" bear that lives in the Yakutat area is probably a color phase of the black bear (*Ursus americanus*).
22. We have a saying in South Carolina, "Thank God for Mississippi," which is ranked fifty-second in everything considered by most to be worthwhile, such as education, quality of life, etc.
23. Hann, R. W., "Follow-up Field Survey of the Oil Pollution from the Tanker *Metula*," report to the U.S. Coast Guard, Research and Development Program, 1975.
24. Smith, J. E., ed., *Torrey Canyon Pollution and Marine Life* (Cambridge: Univ. Print. House, 1968).
25. Hann, R. W., "VLCC *Metula* Oil Spill," Coast Guard Report NTIS no. AS/A-003 805/9WP, 1974.
26. These birds were lesser rheas, also known as Darwin's rheas (*Pterocnemia pennata*).
27. International Petroleum Industry Environmental Conservation Association, *Sensitivity Mapping for Oil Spill Response* (1996).
28. He was talking about the Arctic fox (*Alopex lagopus*). This fox has two phases, blue and white, and the fur historically has been a valuable trade item. Hunt, H. B., and R. P. Grossenheider, *Mammals: R. T. Peterson Field Guides* (Norwalk, Conn.: Eaton Press, 1976).
29. This is the terminal from which the Iraqis released a considerable amount of oil during the Gulf War oil spill of 1991.
30. A rookery is a breeding area for gregarious birds, in this case, sea birds.
31. Wheelwright, Jeff, *Degrees of Disaster* (New York: Simon & Schuster, 1994).
32. Duncan, David James, *The River Why* (New York: Bantam Books, 1984).
33. Maclean, Norman, *A River Runs Through It* (Chicago: University of Chicago Press, 1976).
34. Some people question whether the author actually was a woman.

35. Carcci, Al, and Bob Nastasi, *Hatches II, A Complete Guide to Fishing the Hatches of North American Trout Streams*, (New York: Lyons and Burford, 1995).

36. A sheen is a thin layer of oil on the water surface which generates dull to rainbow colors. Sheens are usually less than 0.1 mm thick, but though thin, sheens can affect animals on the water surface, such as birds.

37. Quinn, Daniel, *Ishmael* (New York: Bantam/Turner Books, 1992).

38. Kathy Sullivan was an astronaut on the space shuttle flight during the *Mt. Mitchell* survey.

39. The reason I know these numbers is that the AAPG required a report from me on the number of attendees at each talk, presumably as input as to whether or not to send anybody else back there.

40. Neither deer nor mountains occur on the North Slope, as far as I know. However, an oil spill seems like a pretty good possibility.

41. Hirschfelder, Arlene (ed.), *Native Heritage* (New York: Macmillan, 1995).

42. Committee on Restoration of Aquatic Ecosystems, *Restoration of Aquatic Ecosystems* (Washington, D.C.: National Academy Press, 1992).

43. Thayer, G. W. (ed.), *Restoring the Nation's Marine Environment* (College Park, Maryland: Maryland Sea Grant Book, 1992).

44. Quoted in Committee on Restoration of Aquatic Ecosystems, *Restoration of Aquatic Ecosystems* (Washington, D.C.: National Academy Press, 1992).

45. Brown, D., *Bury My Heart at Wounded Knee: An Indian History of the American West* (New York: H. Holt and Co., 1970).

Printed and bound by CPI Group (UK) Ltd, Croydon, CR0 4YY

27/10/2024

14580154-0002